Universitext

D0940913

Universitext

Editors (North America): J.H. Ewing, F.W. Gehring, and P.R. Halmos

Asuman G. Aksoy
Mohamed A. Khamsi

Nonstandard Methods in Fixed Point Theory

Springer-Verlag
New York Berlin Heidelberg
London Paris Tokyo Hong Kong

Asuman Güven Aksoy
Department of Mathematics
Claremont McKenna College
Claremont, California 91711
USA

Mohamed A. Khamsi
Department of Mathematical Sciences
University of Texas at El Paso
El Paso, Texas 79968
USA

Mathematical Subject Classification: 47 H10

Library of Congress Cataloging-in-Publication Data
Aksoy, Asuman.
     Nonstandard methods in fixed point theory / Asuman Aksoy, M.A.
Khamsi.
        p.   cm. — (Universitext)
     Includes bibliographical references.
     ISBN 0-387-97364-8
     1. Fixed point theory. 2. Nonexpansive mappings. I. Khamsi,
M.A. II. Title. III. Series.
     QA329.9.A37   1990
515′.7248 — dc20   90-39173

Camera-ready copy provided by the authors.
Printed and bound by R.R. Donnelley & Sons, Harrisonburg, Virginia.
Printed in the United States of America.

9 8 7 6 5 4 3 2 1

ISBN 0-387-97364-8 Springer-Verlag New York Berlin Heidelberg
ISBN 3-540-97364-8 Springer-Verlag Berlin Heidelberg New York

Introduction

Fixed point theory of course entails the search for a combination
of conditions on a set S and a mapping $T : S \to S$ which, in turn,
assures that T leaves at least one point of S fixed, i.e. $x = T(x)$
for some $x \in S$. The theory has several rather well-defined (yet
overlapping) branches. The purely topological theory as well as those
topics which lie on the borderline of topology and functional analysis
(e.g., those related to Leray-Schauder theory) have their roots in the
celebrated theorem of L. E. J. Brouwer. This book is concerned
with another branch of functional analytic theory, a branch which
might be more properly viewed as a far reaching outgrowth of the
contraction mapping principle of Banach and Picard. The mappings
involved are of the form $T : K \to K$, K a subset of a Banach
space, where T is *nonexpansive*; thus $\|T(x) - T(y)\| \leq \|x - y\|$ for
all $x, y \in K$.

Because of its important linkages with the theory of monotone
and accretive operators, fixed point theory for nonexpansive mappings has long been considered a fundamental part of nonlinear functional analysis. However, the attempt to classify those subsets of Banach spaces which have the fixed point property for such mappings
has now become a study in its own right – one which has yielded
many elegant results and led to numerous discoveries in Banach space
geometry. The aim of this book is to give a unified account of the
major new developments inspired by B. Maurey's application in 1980
of Banach space ultraproducts to the theory.

To place these results in perspective we review some history.
Fixed point theory for nonexpansive mappings has its origins in the
1965 existence theorems of F. Browder, D. Gohde, and W. A. Kirk.
Although such mappings are natural extensions of the contraction
mappings, it was clear from the outset that the study of nonexpan-

sive mappings required techniques which go far beyond the purely metric approach. The traditional methods used in studying nonexpansive mappings have involved an intertwining of geometrical and topological arguments. Over the past twenty-five years these methods have yielded substantial results, both of a constructive and nonconstructive nature (the former not requiring the Axiom of Choice). For example, the original theorems of Browder and Gohde exploited the special geometrical structure of uniformly convex Banach spaces. The somewhat more general theorem of Kirk asserts that K has the fixed point property for nonexpansive mappings if K is weakly compact and convex and at the same time possesses a property called 'normal structure' (which means that any convex subset H of K which contains more than one point must contain a point z which has the property:

$$\sup\{\|z - y\| : y \in H\} < \sup\{\|x - y\| : x, y \in H\}).$$

Thus the underlying domain is assumed to possess both topological and geometric properties. At the same time it should be noted that while K is assumed to be compact in the weak topology, T is only norm continuous – thus the Schauder-Tychonoff theorem does not apply. (We also remark that while this theorem was initially established via an application of Zorn's lemma, constructive proofs are now known to exist.)

The early phases of the development of the theory centered around the identification of classes of spaces whose bounded convex sets possess normal structure, and it was soon discovered that certain weakenings and variants of normal structure also suffice. In terms of the classical spaces of functional analysis this approach yielded the following facts: All bounded closed and convex subsets of the ℓ^p and L^p spaces, $1 < p < \infty$, have the fixed point property for nonexpansive mappings, as do all weak* compact convex subsets of ℓ^1 (L. Karlovitz) and all order intervals (closed balls) in ℓ^∞ and L^∞ (R. Sine; P. Soardi). It was shown that the closed convex hull of a weakly convergent sequence in c_0 has this property also (E. Odell and Y. Sternfeld), and that bounded closed convex subsets of certain renormings of ℓ^2 which fail to have normal structure may still have the fixed point property for nonexpansive mappings (L. Karlovitz; J. B. Baillon and R. Schoneberg). A number of more abstract results were also discovered, along with important discoveries related both to the structure of the fixed point sets and to techniques for approximating fixed points. However, in concrete terms, the classical results described above represented the only substantial contributions to the existence part of the theory until 1980 when D. Alspach discovered an example (published in 1981) of a weakly compact convex subset of L^1 which fails to have the fixed point property for nonexpansive mappings. Alspach's example showed that *some* assumption in addition to weak compactness is needed and at the same time it set the stage for Maurey's surprising discovery that all bounded closed convex subsets of *reflexive subspaces* of L^1 do have the fixed point property for nonexpansive mappings. Maurey also showed that the same is true of all weakly compact convex subsets of c_0 and of the Hardy space H^1.

Maurey's methods are nonconstructive in the sense that they involve the Banach space ultraproduct construction. Ultraproduct methods are fundamental to set theory and over the years have found applications in many other branches of mathematics, including both algebra and analysis. However the set-theoretic ultraproduct of Banach spaces in general is not a Banach space and the modification used by Maurey was not put in explicit form until 1972 when it was introduced by D. Dacunha-Castelle and J. Kirvine. Since 1980 Banach space ultraproducts and related ultraproduct methods have yielded many results in fixed point theory for nonexpansive mappings. Also, in those instances where 'standard' techniques have subsequently been shown to yield the same results, the standard approach often appears to be neither simpler nor more intuitive.

Because the methods involved are fairly sophisticated, roughly the first third of this book is devoted to laying a careful foundation for the actual fixed point-theoretic results which follow. The text begins with a careful review of the concepts of Schauder bases, filters, ultrafilters, limits over ultrafilters, and nets. In Chapter 2 both the set-theoretic and Banach space ultraproduct constructions are presented in detail, and finite representability, the Banach-Saks properties, and the ultraproduct of mappings are also discussed. The chapter concludes with a description of the fundamental spaces of Tzirelson and James. The final chapter begins with an introduction to the classical approach to the theory, including a discussion of normal structure. Then, after translating these basic results into ultraproduct language, several new fixed point theorems, including the results of Maurey, are presented. The chapter concludes with a recent application of ultranets due to Kirk.

A major part of the classical fixed point theory for nonexpansive mappings may be found in the recent book: *Topics in Metric Fixed Point Theory* by K. Goebel and W. A. Kirk. However, the nonstandard methods are treated only briefly by Goebel and Kirk; thus the present work might be viewed as complementary to that book. At the same time it should be emphasized that the treatment given here is self-contained in the sense that all results pertinent to the development are included.

—W. A. Kirk

Contents

Chapter 0

SCHAUDER BASES

It is certainly true that the concepts of algebraic bases and orthonormal bases are crucial to the study of finite dimensional vector spaces and Hilbert spaces, respectively. That is why one tries to find a corresponding concept in the study of Banach spaces. Although, using Zorn's lemma, one can always establish the existence of an algebraic basis of a given Banach space, this isn't really sufficient because such a basis does not provide information about the topology of the space.

A sequence (x_n) in a Banach space X is called a *Schauder basis* (or just a *basis*) for X if for each $x \in X$, there exists a unique sequence of scalars (a_n) such that

$$x = \lim_n \sum_{k=1}^n a_k x_k.$$

In this chapter we will give some basic results concerning Schauder bases, generally without proofs. To see the omitted proofs or for more information on Schauder bases, one can consult [16, 53, 143, 169, 179].

It is easy to see that a Schauder basis consists of independent vectors. One important notion for our purposes is a basic sequence:

Definition 0.1. A sequence (x_n) in a Banach space X is called a *basic sequence* if it is a basis for its closed linear span, $\overline{\text{span}}(x_n)$, that is, if for each $x \in \overline{\text{span}}(x_n)$ one can find a unique sequence of scalars (a_n) such that the series $\sum_n a_n x_n$ converges to x. Obviously when

1

$\overline{\text{span}}(x_n) = X$, saying (x_n) is a basic sequence is equivalent to saying (x_n) is a Schauder basis.

We will usually assume that basic sequences (x_n) are normalized, i.e. that $\|x_n\| = 1$ for all $n \in \mathbb{N}$.

The question of whether a Schauder basis for a given Banach space exists has to be carefully considered. First of all, the existence of a Schauder basis implies that the space is separable. Furthermore, although the classical and most common Banach spaces (like ℓ^p $(1 \le p < \infty)$, c_0, and $L^p[0,1]$) all have Schauder bases, there are separable Banach spaces without bases. (See Enflo [62].)

Once it is known that a Banach space has a Schauder basis, it is natural to raise the question of its uniqueness. In order to study this question we need the notion of equivalence between bases. We say that the basic sequences (x_n) in X and (y_n) in Y (where X and Y are Banach spaces) are *equivalent* if $\sum a_n x_n$ converges whenever $\sum a_n y_n$ converges, and vice versa. It follows immediately from the closed graph theorem that (x_n) is equivalent to (y_n) if and only if there exists an isomorphism T from $X_0 = \overline{\text{span}}(x_n)$ onto $Y_0 = \overline{\text{span}}(y_n)$ for which $Tx_n = y_n$ for all n.

It turns out that even up to equivalent bases, if a Schauder basis exists at all it is never unique. In fact, it is known (see [143]) that if X is an infinite dimensional space with at least one Schauder basis, then there are uncountably many mutually non-equivalent normalized bases of X! However, if we perturb each element of a basis by a sufficiently small vector we get a basis which is still equivalent to the original one. A very useful method of doing this uses the notion of block bases.

Definition 0.2 Let (x_n) be a basic sequence in a Banach space X. A sequence of non-zero vectors (u_m) in X of the form $u_m = \displaystyle\sum_{n=p_m+1}^{p_{m+1}} a_n x_n$, where (a_n) are scalars and $p_1 < p_2 < \ldots$ is an increasing sequence of integers, is called a *block basic sequence* or more briefly a *block basis* of (x_n).

Sometimes we will say that a vector u is a *block* of the basic sequence (x_n) if $u = \displaystyle\sum_{i=n_1}^{n_2} a_i x_i$ with $n_1 < n_2$. We also associate to any $x = \sum_n a_n x_n$ a set of integers, called the *support* of x, defined by

$$\text{supp } x = \{n : a_n \neq 0\}. \tag{0.1}$$

The basicity of a sequence (x_n) can be characterized by a condition on the projections on finite subsets of the coordinates.

Theorem 0.3. *A sequence (x_n) is basic if and only if there exists a number $c > 0$ such that for any positive integers n and p and any sequence of scalars (a_i), we have*

$$\left\| \sum_{i=1}^{n} a_i x_i \right\| \leq c \left\| \sum_{i=1}^{n+p} a_i x_i \right\|. \tag{0.2}$$

The least number c that satisfies (0.2) is called the *basis constant* of (x_n). If the basis constant is 1, then $\left(\left\| \sum_i a_i x_i \right\| \right)_{i=1}^{n}$ is a monotone increasing sequence, and the basis is said to be monotone. The inequality (0.2) suggests the following definition.

Definition 0.4. Let (e_n) be a basic sequence in X and write $X_0 = \overline{\text{span}}(e_n)$. A *natural projection* on (e_n) is a mapping $P_F : X_0 \to X_0$ defined by

$$P_F \left(\sum_{i=1}^{\infty} a_i e_i \right) = \sum_{i \in F} a_i e_i, \tag{0.3}$$

where F is a nonempty subset of $\mathbb{N}$.

Clearly P_F is a linear projection onto $\overline{\text{span}}_{n \in F} (e_n)$.

The characterization given in Theorem 0.3 can be formulated as: the sequence (e_n) is basic if and only if the projections $(P_{[0,n]})$ are uniformly bounded. ($[0, n]$ denotes the segment of integers between, and including, 0 and n.) We usually use the notation P_n for $P_{[0,n]}$. The basis consant of (e_n) is therefore given by

$$c = \sup_n \|P_n\|. \tag{0.4}$$

We will say that the basic sequence (e_n) is *bimonotone* if

$$\|P_n\| = \|I - P_n\| = 1 \quad \text{for any } n \in \mathbb{N}.$$

Now let us discuss the relationship between basicity and duality. Without any loss of generality, we will always work with Banach spaces with Schauder bases, rather than using the subspace generated by a basic sequence.

Definition 0.5. Let (e_n) be a Schauder basis for X. For every $n \in \mathbb{N}$, the linear functional e_n^* on X defined by

$$e_n^*\left(\sum_i a_i e_i\right) = a_n \tag{0.5}$$

is bounded. These functionals (e_n^*), which are charaterized by

$$e_n^*(e_m) = \delta_n^m, \tag{0.6}$$

are called the *biorthogonal functionals* associated with the basis (e_n).

It is easy to verify that $\|e_n^*\| \leq 2c$, where c is the basis constant of (e_n). In fact, let (P_n) be the natural projections associated with (e_n); then for every integers $n < m$ we have $P_n^*\left(\sum_{i \leq m} a_i e_i^*\right) = \sum_{i \leq n} a_i e_i^*$. Hence, by Theorem 0.3, the sequence (e_n^*) is a basic sequence in X^* whose basic constant is also c. However, in general (e_n^*) is not a Schauder basis of X^*. Indeed, one can show that the dual of ℓ_1 is the non-separable space ℓ_∞ which therefore cannot have a basis.

Definition 0.6. A Schauder basis (e_n) of X is said to be *shrinking* if (e_n^*) is a Schauder basis of X^*.

As an example of a shrinking Schauder basis, one can take the canonical basis of c_0; the canonical basis of ℓ_1 is not shrinking. But the latter has a surprising property which suggests the following.

Definition 0.7. A *boundedly complete basis* for X is a basis (e_n) for which $\sum_i a_i e_i$ converges whenever the sequence of scalars (a_n) has the property that $(\|\sum_{i=1}^n a_i e_i\|)$ is a bounded sequence.

The natural basis of c_0 is clearly not boundedly complete. As this suggests, there is a kind of duality between shrinkingness and boundedly completeness (see [143]). In the next theorem we give a very useful description of X^{**} when X has a shrinking basis.

Theorem 0.8. *Let (e_n) be a shrinking basis of a Banach space X. Then X^{**} can be identified with the space of all sequences of scalars (a_n) such that $\sup\limits_{n} \| \sum\limits_{i=1}^{n} a_i e_i \| < \infty$. The correspondence is given by*

$$x^{**} \leftrightarrow (x^{**}(e_0^*), x^{**}(e_1^*), x^{**}(e_2^*), \ldots). \qquad (0.7)$$

The norm of x^{**} is equivalent (and, if the basis constant is 1, even equal) to $\sup\limits_{n} \| \sum\limits_{i=1}^{n} a_i e_i \|$.

As a corollary to Theorem 0.8 one can give a new characterization of reflexivity in terms of bases.

Corollary 0.9. *If X has a Schauder basis (e_n), then X is reflexive if and only if (e_n) is both shrinking and boundedly complete.*

Let us remark that the existence of a Schauder basis in a Banach space does not give us very much information about the structure of the space. So if we wish to use bases to study the structure of a Banach space in any detail, we are led to consider bases with various special properties.

Definition 0.10. We say that a Schauder basis (e_n) is *unconditional* if whenever the series $\sum\limits_{i} a_i e_i$ converges, it converges unconditionally, i.e. $\sum\limits_{i} a_{\pi(i)} e_{\pi(i)}$ converges for any permutation π of $\mathbb{N}$.

Proposition 0.11. *For a Schauder basis (e_n), the following are equivalent:*

(1) (e_n) is an unconditional basis;

(2) for every convergent series $\sum\limits_{n} a_n e_n$ and every sequence (ϵ_n) with $\epsilon_n = \pm 1$ for all $n \in \mathbb{N}$, the series $\sum\limits_{n} a_n \epsilon_n e_n$ converges;

(3) for every convergent series $\sum_n a_n e_n$ and every sequence of scalars (b_n) such that, for all n, $|b_n| \le |a_n|$, the series $\sum_n b_n e_n$ converges;

(4) there exists $c \ge 1$ such that if A and B are finite subsets of $\mathbb{N}$ with $A \subset B$, then for any sequence (a_n) of scalars

$$\left\| \sum_{n \in A} a_n e_n \right\| \le c \left\| \sum_{n \in B} a_n e_n \right\| \qquad (0.8)$$

(we shall call the smallest c satisfying (0.8) the unconditional basis constant *of (e_n), and the basis will be called* unconditionally monotone *if $c = 1$);*

(5) For every convergent series $\sum_n a_n e_n$ and every strictly increasing sequence of integers (n_i), the series $\sum_i a_{n_i} e_{n_i}$ converges.

Some bases have a nice property similar to the one described immediately above.

Definition 0.12. A basic sequence (e_n) is called *spreading* if for every convergent series $\sum_i a_i e_i$ and every strictly increasing sequence of integers (n_i), the series $\sum_i a_i e_{n_i}$ converges and

$$\left\| \sum_i a_i e_i \right\| = \left\| \sum_i a_i e_{n_i} \right\|. \qquad (0.9)$$

Examples of such bases will be given in section II of Chapter 2. The uniform boundedness principle can be used to prove that (e_n) is an unconditional basis whenever there exists a constant $\lambda \ge 1$ such that for every convergent series $\sum a_i e_i$ and every sequence of signs (ϵ_n) $(\epsilon_n = \pm 1)$, the series $\sum a_i \epsilon_i e_i$ converges and satisfies

$$\left\| \sum a_i \epsilon_i e_i \right\| \le \lambda \left\| \sum a_i e_i \right\|. \qquad (0.10)$$

The smallest constant λ which satisfies (0.10) is called the *unconditional constant* of (e_n). We will say later that the sequence (e_n) is

λ-unconditional, referring to the unconditional constant λ. It is not hard to show that

$$c \leq \lambda \leq 2c. \qquad (0.11)$$

Let us add that Pelczynski [155] (see also [151]) gave a simple proof of the fact that $L^1[0,1]$ cannot be isomorphic to a subspace of a space with an unconditional basis. One deep open problem in Banach space theory is to determine whether any infinite dimensional Banach space contains an unconditional basic sequence. Enflo [62] gave an example of a separable Banach space which fails to have a basis, while it is easy to prove that any Banach space contains a basic sequence (see [143], p. 4).

The conclusion of Corollary 0.9 can be strengthened when unconditional Schauder bases are involved.

Theorem 0.13. *Let X be a subspace of a space with an unconditional Schauder basis. Then X is reflexive unless it contains a subspace isomorphic to either c_0 or ℓ_1.*

Let us give a proof of one technical result which will be very useful throughout the rest of this book.

Proposition 0.14. *Let (e_n) be a normalized Schauder basis of X with associated biorthogonal system (e_n^*). Let (x_k) be a bounded sequence such that $(e_n^*(x_k)) \to 0$ as $k \to \infty$, i.e. the coefficients of (x_k) converge pointwise to 0. Then there is a subsequence (x_{k_i}) of (x_k) and a sequence (u_i) of succesive blocks of (e_n) such that*

$$\lim_i \|x_{k_i} - u_i\| = 0.$$

Proof. Let (ϵ_i) be a sequence of positive numbers going to 0. One can find $N_0 \in \mathbf{N}$ such that $\|x_{N_0} - P_{N_0} x_{N_0}\| \leq \epsilon_0$. Since $\|P_{N_0} x_n\| \to 0$ as $n \to \infty$, one can find $n_1 > N_0$ such that $\|P_{N_0} x_n\| \leq \epsilon_1$ for any $n \geq n_1$. Then let $N_1 > N_0$ satisfy $\|x_{n_1} - P_{N_1}(x_{n_1})\| \leq \epsilon_1$. Since $\|P_{N_1} x_n\| \to 0$ as $n \to \infty$, one can find $n_2 > n_1$ such that $\|P_{N_1} x_n\| \leq \epsilon_2$ for $n \geq n_2$. Then let $N_2 > N_1$ be such that $\|P_{N_2} x_{n_2} - x_{n_2}\| \leq \epsilon_2$. We are constructing a sequence of pairs $((n_i, N_i))$ with $n_1 < n_2 < \ldots$ and $N_1 < N_2 < \ldots$, and which satisfy

$$\|P_{N_i}(x_n)\| \leq \epsilon_k \text{ for } n \geq n_k$$

and
$$\|P_{N_k} x_{n_k} - x_{n_k}\| \le \epsilon_k.$$

Suppose that the first k pairs have been constructed; then since $\|P_{N_k} x_n\| \to 0$ as $n \to \infty$, one can find $n_{k+1} > n_k$ such that $\|P_{N_k} x_n\| \le \epsilon_{k+1}$ for $n \ge n_{k+1}$, and then we can find $N_{k+1} > N_k$ such that $\|P_{N_{k+1}} x_{n_{k+1}} - x_{n_{k+1}}\| \le \epsilon_{k+1}$. So the process indeed is inductively sound. If we put $u_k = (I - P_{N_k} + P_{N_{k-1}})x_{n_k}$ for $k = 1, 2, \ldots$, we obtain
$$\|x_{n_k} - u_k\| \le \epsilon_{k-1} + \epsilon_k.$$

The support of u_k is clearly in the interval $[N_{k-1}, N_k]$. This completes the proof. ∎

We will complete this chapter by showing a way to generalize the notion of a Schauder basis to a wide class of Banach spaces. Note that a Schauder basis in a sense decomposes a Banach space into a sum of one-dimensional spaces. It is sometimes useful to consider cruder decompositions, where the components into which we decompose the space are of dimension larger than 1.

Definition 0.14. Let X be a Banach space. A sequence (x_n) of closed subspaces of X is called a *Schauder decomposition* of X if every $x \in X$ has a unique representation of the form $x = \sum_n x_n$ with $x_n \in X_n$ for all n.

Observe that if $\dim X_n = 1$ for every n, then a Schauder decomposition is no different from a Schauder basis. The decompositions which are most useful in applications are those in which $\dim X_n < \infty$ for all n (sup$_n$ $\dim X_n$ need not be finite). Such decompositions are called *finite dimensional decompositions* or *F.D.D.'s* for short. Many results concerning bases generalize trivially to Schauder decompositions.

One can ask whether there is a close relationship between the existence of F.D.D.'s and the existence of bases. By Enflo's example [62], we know there are separable Banach spaces which fail to have an F.D.D.

Let us mention that in ([143], p. 51), one can find an example of a Banach space which has an unconditional F.D.D. However, the space

8

does not have an unconditional basis and it is not even complemented in a space with an unconditional basis.

Chapter 1

I. FILTERS

We will begin this chapter by defining filters and ultrafilters and giving some important results concerning ultrafilters. For more details the reader may consult [25], [98].

Definition 1.1. Let I be a non-empty set. A *filter* $\mathcal{F}$ on I is a nonempty collection of subsets of I satisfying

(1.1) if $A, B \in \mathcal{F}$ then $A \cap B \in \mathcal{F}$;
(1.2) if $A \in \mathcal{F}$ and $A \subset B$ then $B \in \mathcal{F}$.

It is clear from the above definition that I is an element of any filter on I. The following are some examples of filters to which we will refer throughout the rest of the text.

Examples 1.2.

(a) Let $i \in I$ be fixed and consider $\mathcal{F}_i = \{A \subset I : i \in A\}$. $\mathcal{F}_i$ is a filter on I and is called a *trivial filter*.

(b) Define $\mathcal{F} = \{A \subset I : I - A \text{ is finite}\}$. This is also a filter on I and is called the *Fréchet filter*.

(c) Let I be a topological space with $i \in I$, and let $\mathcal{F}(i) = \{V : V$ is a neighborhood of $i\}$. Then $\mathcal{F}(i)$ is a filter on I associated with i.

(d) Let $(I, \leq)$ be an ordered set. Define

$$\mathcal{F} = \{B \subset I : \text{ there exists } i_0 \in I \text{ such that } i \in B \text{ for any } i \geq i_0\}.$$

$\mathcal{F}$ is a filter on I provided that I is *upward directed*, i.e. for any $i, j \in I$ there exists $k \in I$ such that $i \leq k$ and $j \leq k$.

Proposition 1.3. *Let I be a set and suppose $\mathcal{B} \subset 2^I$ is stable under intersection. Define*

$$\mathcal{F}(\mathcal{B}) = \{A \subset I : \text{ there is a } B \in \mathcal{B} \text{ such that } B \subset A\}.$$

Then $\mathcal{F}(\mathcal{B})$ is a filter on I. Moreover, any filter which contains $\mathcal{B}$ must contain $\mathcal{F}(\mathcal{B})$.

If $\mathcal{F}$ is a filter on I and $\phi \in \mathcal{F}$, then $\mathcal{F} = 2^I$ and is called the *improper filter*. It should be clear that $\mathcal{F}$ is a trivial filter if and only if $\phi \notin \mathcal{F}$ and there exists an $i \in I$ such that $\{i\} \in \mathcal{F}$.

Since any filter on I contains I, the intersection of any family of filters on I is nonempty; and one can easily show that such an intersection is itself a filter on I. However, we don't have a similar result for the union of a family of filters on I except when the family is upward directed.

Now let $\mathcal{P}$ be the family of all proper filters on I,

$$\mathcal{P} = \{\mathcal{F} : \mathcal{F} \text{ is a filter on } I, \mathcal{F} \neq 2^I\}.$$

That is, $\mathcal{F} \in \mathcal{P}$ if $\mathcal{F}$ is a filter on I and $\phi \notin \mathcal{F}$. Since $\mathcal{P}$ is an *inductive* set (every increasing chain has an upper bound), it follows from Zorn's lemma that $\mathcal{P}$ has a maximal element. In other words, there exists some $\mathcal{F} \in \mathcal{P}$ such that if $\mathcal{D} \in \mathcal{P}$ and $\mathcal{F} \subset \mathcal{D}$, then $\mathcal{F} = \mathcal{D}$.

Definition 1.4. The maximal elements of $\mathcal{P}$ are called *ultrafilters* on I.

In general there is no easy characterization of the maximal elements of an ordered set, since their existence depends on Zorn's lemma and not on a constructive proof. However, for ultrafilters we have the following proposition.

Proposition 1.5. *A filter $\mathcal{U}$ on I is an ultrafilter if and only if*

(1.3) *for every $A \subset I$, either A or $I - A$ belongs to $\mathcal{U}$.*

Proof. Assume that $\mathcal{U}$ satisfies (1.3) and let $\mathcal{F}$ be a filter on I such that $\mathcal{U} \subset \mathcal{F}$ and $\mathcal{U} \neq \mathcal{F}$. Pick $A \in \mathcal{F} - \mathcal{U}$; then condition (1.3) implies that $I - A \in \mathcal{U}$. Since $\mathcal{U} \subset \mathcal{F}$, we deduce that A and $I - A$

are in $\mathcal{F}$, which in turn implies $\phi = A \cap (I - A) \in \mathcal{F}$. Therefore we have $\mathcal{F} = 2^I$, so that $\mathcal{U}$ is maximal, i.e. an ultrafilter.

Conversely, suppose that $\mathcal{U}$ is an ultrafilter on I and let $A \subset I$; assume that $I - A \notin \mathcal{U}$; we will show that $A \in \mathcal{U}$. Consider $B = \{A \cap B : B \in \mathcal{U}\}$. Let $\mathcal{F}(B)$ be the filter on I generated by B, as defined in Proposition 1.3. First of all, we have $\mathcal{U} \subset \mathcal{F}(B)$: if $B \in \mathcal{U}$, then $A \cap B \in B \subset \mathcal{F}(B)$, so by (1.2) $B \in \mathcal{F}(B)$. We will complete the proof by showing that $\mathcal{U} = \mathcal{F}(B)$. Now because $I - A \notin \mathcal{U}$, it follows that $\phi \notin B$ (again using (1.2)); hence $\mathcal{F}(B)$ is a proper filter, and the maximality of $\mathcal{U}$ implies that $\mathcal{U} = \mathcal{F}(B)$. Since $A \in \mathcal{F}(B)$, we have the desired conclusion. ∎

Remarks 1.6.

(1) Using Zorn's lemma suitably, we can extend any proper filter on I to an ultrafilter on I.

(2) Any trivial filter on I is an ultrafilter.

Proposition 1.5 yields the following result, which will be crucial to our discussion.

Proposition 1.7. *Let $\mathcal{U}$ be an ultrafilter on I. Assume that $I_1 \cup I_2 \cup \cdots \cup I_n \in \mathcal{U}$, where each $I_i \subset I$. Then for some k, $I_k \in \mathcal{U}$.*

Proof. To the contrary, assume that $I_k \notin \mathcal{U}$ for each $k = 1, 2, \ldots, n$. Then by Proposition 1.5 we have $I - I_k \in \mathcal{U}$. Therefore $\phi = \bigcap_{1 \leq k \leq n} (I - I_k) \cap (\bigcup_k I_k) \in \mathcal{U}$, which is a contradiction. ∎

Using Proposition 1.7, we can state another characterization of trivial filters.

Proposition 1.8. *The ultrafilter $\mathcal{U}$ on I is trivial if and only if there exists a finite set $A \in \mathcal{U}$.*

Throughout these notes we will use ultrafilters on $\mathbf{N}$. In this case the ultrafilters have a surprising property.

Proposition 1.9. *Every nontrivial ultrafilter $\mathcal{U}$ on $\mathbf{N}$ is countably incomplete, i.e. there exists a sequence (A_n) with each $A_n \in \mathcal{U}$ and with $\bigcap_n A_n = \phi$.*

Proof. Let $n \in \mathbb{N}$. Our assumption on $\mathcal{U}$ implies that there exists an $A_n \in \mathcal{U}$ such that $n \notin A_n$. (In fact, we may pick $A_n = \mathbb{N} - \{n\}$.) Clearly, we have $\bigcap_n A_n = \phi$. ∎

II. LIMITS OVER FILTERS

In this section, we consider a Hausdorff topological space X. We will first define convergence in X with respect to a filter or ultrafilter; this will lead to several important results.

Definition 2.1. Let $(x_i)_{i \in I}$ be a collection of elements of X indexed by a set I, and consider a filter $\mathcal{F}$ on I. We say that (x_i) *converges to* $x \in X$ *over* $\mathcal{F}$ if the set $\{i \in I : x_i \in V\}$ is in $\mathcal{F}$ for any neighborhood V of x. The limit will be denoted by $\lim_{i,\mathcal{F}} x_i$ or $\lim_{\mathcal{F}} x_i$.

Notice that when $\mathcal{F}$ is proper, the limit over $\mathcal{F}$ is unique. If, moreover, C is a closed subset of X and $\{x_i\} \subset C$, then $\lim_{\mathcal{F}} x_i$ belongs to C.

If $\mathcal{F}_{i_0}$ is the trivial filter generated by $i_0 \in I$, then $\lim_{\mathcal{F}_{i_0}} x_i = x_{i_0}$. Trivial filters give us no information on asymptotic behavior of sets, so we will generally avoid them.

Proposition 2.2. *Let $\mathcal{U}$ be a nontrivial ultrafilter on $\mathbb{N}$ and suppose (x_n) converges to x in the topology of the space X. Then (x_n) converges to x with respect to the filter $\mathcal{U}$, i.e. $\lim_{\mathcal{U}} x_n = x$.*

Proof. Let V be any neighborhood of x. Since the set $\{i : x_i \notin V\}$ is finite, Propositions 1.5 and 1.8 imply that $\{i : x_i \in V\}$ is in $\mathcal{U}$. ∎

Remark 2.3. Let X be a metric space. If $\mathcal{U}$ is a nontrivial ultrafilter and $\lim_{\mathcal{U}} x_n = x$ then there exists a subsequence of (x_n) which converges to x with respect to the topology of X.

The next result is interesting because it shows how ultrafilters can be used to characterize compactness of a topological space. One can see this result as a generalization of the well-known characterization of compactness in metric spaces.

Theorem 2.4. *Let K be a Hausdorff topological space. K is compact if and only if any set $(x_i)_{i \in I} \subset K$ is convergent over any ultrafilter $\mathcal{U}$ on I.*

Proof. Assume K is compact. Let $(x_i)_{i \in I}$ be a subset of K and let $\mathcal{U}$ be an ultrafilter on I, and suppose that $(x_i)_{i \in I}$ does not converge to any $x \in K$. Then each x in K has a neighborhood V_x such that $\{i \in I : x_i \in V_x\} \notin \mathcal{U}$. And since $K \subset \bigcup_{x \in K} V_x$ and K is compact, there exist $V_{y_1}, V_{y_2}, \ldots, V_{y_n}$ such that $K \subset \bigcup_{j=1}^{n} V_{y_j}$. This implies that $I = \bigcup_{j=1}^{n} I_j$ where $I_j = \{i \in I : x_i \in V_{y_j}\}$. According to Proposition 1.7, then, some $I_k \in \mathcal{U}$, and we have a contradiction.

Conversely, suppose that any subset $(x_i)_{i \in I}$ of K is convergent over any ultrafilter $\mathcal{U}$ on I. Let $(F_\alpha)_{\alpha \in \Gamma}$ be a family of closed subsets of K which has the finite intersection property. We will prove that $\bigcap_\alpha F_\alpha \neq \phi$. Consider $I = \{A \subset \Gamma : A \text{ is finite}\}$ and let $x_A \in \bigcap_{\alpha \in A} F_\alpha$. Set $\mathcal{B} = \{[A, \infty) : A \in I\} \subset 2^I$, where by $[A, \infty)$ we mean the set $\{B \in I : A \subset B\}$. Since $[A, \infty) \cap [A', \infty) = [A \cup A', \infty)$, $\mathcal{B}$ is stable under intersection, and we may define $\mathcal{F}(\mathcal{B})$ as in Proposition 1.3. $\mathcal{F}(\mathcal{B})$ is a proper filter on I because $\phi \notin \mathcal{B}$, so let $\mathcal{U}$ be some ultrafilter on I which extends $\mathcal{F}(\mathcal{B})$. Now we can use the assumption on K to deduce that $\lim_{A,\mathcal{U}} x_A = x$ exists.

We complete the proof by showing that x is an element of every F_α. Suppose not; say $x \notin F_{\alpha'}$. Then x has a neighborhood V_x such that $V_x \cap F_{\alpha'} = \phi$. Since $\lim_{\mathcal{U}} x_A = x$, we have $I_x = \{A \in I : x_A \in V_x\} \in \mathcal{U}$. We also know $I_x \cap [\{\alpha'\}, \infty) \in \mathcal{U}$, using $[\{\alpha'\}, \infty) \in \mathcal{B} \subset \mathcal{F}(\mathcal{B}) \subset \mathcal{U}$. For any $A \in I_x \cap [\{\alpha'\}, \infty)$, both $x_A \in V_x$ and $x_A \in \bigcap_{\gamma \in A} F_\gamma \subset F_{\alpha'}$, which contradicts our assumption that $V_x \cap F_{\alpha'} = \phi$; hence $I_x \cap [\{\alpha'\}, \infty) = \phi \in \mathcal{U}$. But this is also a contradiction, and the proof is done. ∎

Since we can define limits over filters in linear topological spaces, it is natural to ask if this notion cooperates well with the linear structure. The next proposition answers this question.

Proposition 2.5. *Let X be a linear topological space and $\mathcal{U}$ an ultra-filter on a set I. Suppose that $(x_i)_{i \in I}$ and $(y_i)_{i \in I}$ are two subsets of X and $\lim_{\mathcal{U}} x_i = x$ and $\lim_{\mathcal{U}} y_i = y$ both exist. Then*

$$\lim_{\mathcal{U}}(x_i + y_i) = x + y \quad and \quad \lim_{\mathcal{U}} \alpha x_i = \alpha x$$

for any scalar α.

Proposition 2.6. *Let X be a Banach lattice and $(x_i)_{i \in I}$ a subset of X with each $x_i \geq 0$. Then the limit of $(x_i)_{i \in I}$ over any ultrafilter on I is also positive.*

Proof. This follows from the fact that the positive cone X_+ of X is closed. ∎

We conclude this section with the following result, which has an easy proof.

Proposition 2.7. *Let X and Y be two Hausdorff spaces and let f be a continuous map from X into Y. Also let $(x_i)_{i \in I}$ be in X, and assume that $\lim_{\mathcal{U}} x_i = x$ exists in X (where $\mathcal{U}$ is an arbitrary ultrafilter on I). Then $\lim_{\mathcal{U}} f(x_i)$ exists in Y and is equal to $f(x)$.*

III. NETS

We will now discuss some basic properties of nets and ultranets. These concepts will be important in Chapter 3 when we study certain fixed point theorems.

The notions of nets and ultranets are very useful when an approach based on ordinary sequences fails; this can happen if one is dealing with topologial spaces which are not "of a sequential nature." For a more detailed study of nets and ultranets we refer the reader to [98].

Definition 3.1. A binary relation $\geq$ is said to *direct* the set $\mathcal{D}$ if $\mathcal{D}$ is nonempty and

    1) if $m, n, p \in \mathcal{D}$ such that $m \geq n$ and $n \geq p$, then $m \geq p$;

2) if $m \in \mathcal{D}$ then $m \geq m$;

3) if $m, n \in \mathcal{D}$, then there is a $p \in \mathcal{D}$ such that $p \geq m$ and $p \geq n$.

The pair $(\mathcal{D}, \geq)$ is called a *directed set*. If, in addition to the above properties, for each $m, n \in \mathcal{D}$ either $m \geq n$ or $n \geq m$, then the pair $(\mathcal{D}, \geq)$ is called a *linearly directed set*.

Definition 3.2. Let $\mathcal{D}$ be a directed set and S an arbitrary set. Any mapping $x : \mathcal{D} \to S$ is called a *net*. We will use the notation $\{x_n : n \in \mathcal{D}\}$ to denote a net, where by x_n we mean $x(n)$.

The net $\{x_n : n \in \mathcal{D}\}$ is said to be *eventually* in $G \subset S$ if there exists an $n_0 \in \mathcal{D}$ such that $x_n \in G$ whenever $n \geq n_0$; it is *frequently* in G if for any $n \in \mathcal{D}$ there is an $m \in \mathcal{D}$, $m \geq n$, with $x_m \in G$.

Now if the net $\{x_n : n \in \mathcal{D}\}$ is frequently in G and we set $\mathcal{E} = \{n \in \mathcal{D} : x_n \in G\}$, then $\mathcal{E}$ has the property that for any $n \in \mathcal{D}$ there exists an $m \in \mathcal{E}$ with $m \geq n$; such subsets of $\mathcal{D}$ are called *cofinal*. Note that each cofinal subset of $\mathcal{D}$ is also directed by $\geq$.

Definition 3.3. Let S be a Hausdorff topological space. We say that the net $\{x_n : n \in \mathcal{D}\}$ is *convergent* to $p \in S$ if $\{x_n : n \in \mathcal{D}\}$ is eventually in any neighborhood of p. In this case we write

$$\lim_{n \in \mathcal{D}} x_n = p.$$

Let us show that the limit point p is unique. Assume to the contrary that $\{x_n : n \in \mathcal{D}\}$ converges to both p and p', with $p \neq p'$. Since S is Hausdorff, there exist neighborhoods V_p and $V_{p'}$ of p and p' such that $V_p \cap V_{p'} = \phi$. But $\{x_n : n \in \mathcal{D}\}$ is eventually in both V_p and $V_{p'}$; so pick $m, m' \in \mathcal{D}$ with $x_m \in V_p$ and $x_{m'} \in V_{p'}$. Because $\mathcal{D}$ is directed by $\geq$, there exists $k \in \mathcal{D}$ with $k \geq m$ and $k \geq m'$. It follows that $x_k \in V_p$ and $x_k \in V_{p'}$, which yields a contradiction.

Although the concepts we are developing are a natural generalization of sequential limits, there are major differences between the two cases. Recall that a subsequence of a sequence $\{x_n : n \in \mathsf{N}\}$ is any sequence of the form $\{x_{\varphi(n)} : n \in \mathsf{N}\}$, where $\varphi : \mathsf{N} \to \mathsf{N}$ is an increasing mapping. We will now define the notion of a subnet.

Definition 3.4. A net $\{z_n : n \in \mathcal{D}\}$ is called a *subnet* of the net $\{x_n : n \in \mathcal{E}\}$ if and only if there is a function $\varphi : \mathcal{D} \to \mathcal{E}$ with

1) $z_n = x_{\varphi(n)}$, and

2) for each $m \in \mathcal{E}$, there exists an $n \in \mathcal{D}$ such that $p \geq n$ implies $\varphi(p) \geq m$.

Condition 2) states that when p becomes large, so does $\varphi(p)$. This implies that if a net is eventually in G, then any subnet is also eventually in G. Notice that there is no assumption that the order involved is linear; this fact provides for a very rich system of subnets of any given net.

The next theorem shows how the machinery of nets and subnets can be used to give a characterization of compactness.

Theorem *A Hausdorff topological space X is compact if and only if each net in X has a subnet which converges to a point of X.*

We will postpone discussion of this theorem until after we have introduced the notion of an ultranet.

Definition 3.5. A net $\{x_n : n \in \mathcal{D}\}$ in X is called an *ultranet* or a *universal net* if, given any $G \subset X$, $\{x_n : n \in \mathcal{D}\}$ is either eventually in G or eventually in the complement of G.

As in the case of filters and ultrafilters, a natural question to ask is whether every net yields an ultranet. The following theorem will answer this question and give some basic properties of ultranets.

Theorem 3.6. *Let $\{x_n : n \in \mathcal{D}\}$ be a net in a Hausdorff topological space X. Then*

1) $\{x_n : n \in \mathcal{D}\}$ has a subnet which is an ultranet;

2) if $\{x_n : n \in \mathcal{D}\}$ is an ultranet in X then $\{f(x_n) : n \in \mathcal{D}\}$ is an ultranet in Y, for any continuous $f : X \to Y$;

3) X is compact if and only if every ultranet in X converges.

Proof. We will give a proof for 1). Parts 2) and 3) are easy to verify, using the same techniques as those developed in the ultrafilter context.

III. Nets

Let $\{x_n : n \in \mathcal{D}\}$ be a net in X, and consider

$$\mathcal{F} = \{G \subset \mathcal{D} : \{n : n \in \mathcal{D}\} \text{ is eventually in } G\}.$$

$\mathcal{F}$ is not empty since $\mathcal{D} \in \mathcal{F}$, and it is clear that $\mathcal{F}$ is a filter on $\mathcal{D}$. Thus $\mathcal{F}$ is contained in an ultrafilter $\mathcal{U}$ on $\mathcal{D}$. Define U by

$$\mathsf{U} = \{(n, F) : n \in F \text{ and } F \in \mathcal{U}\}$$

and let U be ordered by $\geq$, where

$$(n, F) \geq (m, G) \text{ if } F \subset G.$$

Also define $\varphi : \mathsf{U} \to \mathcal{D}$ by $\varphi(n, F) = n$. We claim that $\{x_{\varphi(\alpha)} : \alpha \in \mathsf{U}\}$ is a subnet of $\{x_n : n \in \mathcal{D}\}$ and is also an ultranet. To prove the former claim, let $m \in \mathcal{D}$ and put $A = \{n \in \mathcal{D} : n \geq m\}$; then $A \in \mathcal{F} \subset \mathcal{U}$ and $(m, A) \in \mathsf{U}$, since $m \in A$. Now if $(p, G) \geq (m, A)$ then $G \subset A$, and since $p \in G$ we have $p \in A$, i.e. $p \geq m$; therefore $\varphi(p, G) \geq m$. This proves that $\{x_{\varphi(\alpha)} : \alpha \in \mathsf{U}\}$ is a subnet of $\{x_n : n \in \mathcal{D}\}$. The maximality of $\mathcal{U}$ implies that $\{x_{\varphi(\alpha)} : \alpha \in \mathsf{U}\}$ is an ultranet. ∎

Chapter 2

I. THE SET-THEORETIC ULTRAPRODUCT

Let $(A_i)_{i \in I}$ be a family of sets and let $\mathcal{U}$ be an ultrafilter on I. By $\prod\limits_{i \in I} A_i$ we mean the cartesian product of the sets $(A_i)_{i \in I}$; consider the relation $\sim_{\mathcal{U}}$ on $\prod\limits_{i \in I} A_i$ defined by:

$$(a_i) \sim_{\mathcal{U}} (b_i) \text{ if and only if } \{i : a_i = b_i\} \in \mathcal{U}. \qquad (1.1)$$

It is easy to show that the relation $\sim_{\mathcal{U}}$ is an equivalence relation on $\prod\limits_i A_i$. We will use $\widetilde{(a_i)}$ to denote the equivalence class of (a_i).

Definition 1.1. The *set-theoretic ultraproduct* of the family $(A_i)_{i \in I}$, denoted by $(A_i)_{\mathcal{U}}$, is defined as the quotient of $\prod\limits_i A_i$ over $\sim_{\mathcal{U}}$. When $A_i = A$ for all $i \in I$, $(A_i)_{\mathcal{U}} = (A)_{\mathcal{U}}$ is called the *set-theoretic ultra-power* of A.

We will simply say "ultraproduct" instead of "set-theoretic ultraproduct" if the context is clear. ("Banach space ultraproducts" will be discussed in section II.) Also, if $A_i \subset B_i$ for all $i \in I$, we will identify $(A_i)_{\mathcal{U}}$ with a subset of $(B_i)_{\mathcal{U}}$ in the obvious way.

Proposition 1.2. *If $(A_i)_{i \in I}$ and $(B_i)_{i \in I}$ are two families of sets, then the following are true:*
  (i) $(A_i)_{\mathcal{U}} \cup (B_i)_{\mathcal{U}} = (A_i \cup B_i)_{\mathcal{U}}$;
  (ii) $(A_i)_{\mathcal{U}} \cap (B_i)_{\mathcal{U}} = (A_i \cap B_i)_{\mathcal{U}}$;

(iii) $(A_i)_{\mathcal{U}} - (B_i)_{\mathcal{U}} = (A_i - B_i)_{\mathcal{U}}$.

Proof. Proofs of *(i)*, *(ii)*, and *(iii)* are similar, so we only give a proof of *(i)*. It is clear that $(A_i)_{\mathcal{U}} \cup (B_i)_{\mathcal{U}} \subset (A_i \cup B_i)_{\mathcal{U}}$. To prove the other inclusion let us take a representative (x_i) of the class $\widetilde{x} \in (A_i \cup B_i)_{\mathcal{U}}$. Consider the sets

$$I_A = \{i : x_i \in A_i\} \text{ and } I_B = \{i : x_i \in B_i\},$$

which satisfy $I_A \cup I_B = I$. By Proposition 1.7 of Chapter 1, we deduce that $I_A \in \mathcal{U}$ or $I_B \in \mathcal{U}$; let us assume that $I_A \in \mathcal{U}$. Define (a_i) so that when $i \in I_A$, $a_i = x_i$, and when $i \notin I_A$, a_i is an arbitrary member of A_i. Then clearly $(a_i) \sim_{\mathcal{U}} (x_i)$, so that $\widetilde{(a_i)}_{\mathcal{U}} = \widetilde{x}$. So $\widetilde{x} \in (A_i)_{\mathcal{U}}$ and we may conclude that $(A_i \cup B_i)_{\mathcal{U}} \subset (A_i)_{\mathcal{U}} \cup (B_i)_{\mathcal{U}}$. ■

There is a natural topology on the ultraproduct of sets with topologies. Let $(K_i)_{i \in I}$ be a family of Hausdorff spaces, and define

$$\mathcal{O} = \left\{ (A_i)_{\mathcal{U}} \subset (K_i)_{\mathcal{U}} : A_i \text{ is an open subset of } K_i \right\}. \qquad (1.2)$$

It is clear from Proposition 1.2 that $\mathcal{O}$ is a basis of a topology on $(K_i)_{\mathcal{U}}$. We will assume that $(K_i)_{\mathcal{U}}$ is always equipped with this topology.

One can consider the same problem when the sets in the ultraproduct are measure spaces $(\Omega_i, \mathcal{A}_i, \mu_i)_{i \in I}$. Without any loss of generality, we will restrict our attention to probability measures. Consider the following collection of subsets of $(\Omega_i)_{\mathcal{U}}$:

$$\widetilde{\mathcal{A}}_0 = \left\{ (A_i)_{\mathcal{U}} : A_i \in \mathcal{A}_i \text{ for } i \in I \right\}. \qquad (1.3)$$

It is easy to verify that $\widetilde{\mathcal{A}}_0$ is a Boolean algebra on $(\Omega_i)_{\mathcal{U}}$. We define a measure $\widetilde{\mu}_0$ on $\widetilde{\mathcal{A}}_0$ by setting

$$\widetilde{\mu}_0(\widetilde{A}) = \lim_{\mathcal{U}} \mu_i(A_i), \qquad (1.4)$$

where $\widetilde{A} = (A_i)_{\mathcal{U}} \in \widetilde{\mathcal{A}}_0$. It follows from Proposition 1.2 that $\widetilde{\mu}_0$ is well-defined and additive. To extend $\widetilde{\mu}_0$ we need the following

Proposition 1.3. *The measure $\widetilde{\mu}_0$ is σ-additive on $\widetilde{\mathcal{A}}_0$. Consequently, $\widetilde{\mu}_0$ can be uniquely extended to a σ-additive measure $\widetilde{\mu}$ on $\widetilde{\mathcal{A}}$, the least σ-algebra containing $\widetilde{\mathcal{A}}_0$.*

Proof. Let $\tilde{A}, \tilde{A}_k \in \tilde{\mathcal{A}}_0$ $(k = 1, 2, \ldots)$ with $\tilde{A}_j \cap \tilde{A}_k = \phi$ for $j \neq k$ and $\tilde{A} = \bigcup_k \tilde{A}_k$. Since $\tilde{\mu}_0$ is additive, we have only to show

$$\tilde{\mu}_0(\tilde{A}) \leq \sum_{k=1}^{\infty} \tilde{\mu}_0(\tilde{A}_k) + \epsilon \tag{1.5}$$

for any $\epsilon > 0$. Let $\tilde{A}_k = (A_i^k)_u$. There exists $I_0 \in \mathcal{U}$ such that $\mu_i(A_i^k) \leq \tilde{\mu}_0(\tilde{A}_k) + \frac{\epsilon}{2^k}$ for all $i \in I_0$. Put $B_i^k = A_i^k$ if $i \in I_0$ and $B_i^k = \phi$ otherwise; clearly $(B_i^k)_{i \in I}$ is another representative of $\tilde{A}_k$. Furthermore we have

$$\mu_i\left(\bigcup_{k=1}^{\infty} B_i^k\right) \leq \sum_{k=1}^{\infty} \tilde{\mu}_0(\tilde{A}_k) + \epsilon$$

for all $i \in I$. Consequently,

$$\tilde{\mu}_0\left(\left(\bigcup_{k=1}^{\infty} B_i^k\right)_u\right) \leq \sum_{k=1}^{\infty} \tilde{\mu}_0(\tilde{A}_k) + \epsilon.$$

But $\tilde{A} = \bigcup_{k=1}^{\infty} \tilde{A}_k \subset \left(\bigcup_{k=1}^{\infty} B_i^k\right)_u$, from which inequality (1.5) follows. $\blacksquare$

The measure space we have obtained will be denoted by $(\Omega_i, \mathcal{A}_i, \mu_i)_u$.

II. THE BANACH SPACE ULTRAPRODUCT

Although the ultraproduct construction was initially a fundamental method of model theory (see [48, 85, 145]), it has influenced several other branches of mathematics, such as algebra and set theory. The step into Banach space theory was motivated by the development of the local theory of Banach spaces, which goes back to the work of Lindenstrauss and Pelczynski [141], Lindenstrauss and Rosenthal [142], and James [89]. The explicit definition of ultraproducts of Banach spaces was introduced by Dacunha-Castelle and Krivine [52] (see also [51]). A more general, yet closely related concept – the nonstandard hull of a Banach space – was developed

II. The Banach Space Ultraproduct

by Luxemburg [145] (see also [85]). For more on ultraproduct construction the reader can consult [84, 174].

Let $(X_i)_{i \in I}$ be a family of Banach spaces. $\ell_\infty(X_i)$ will denote the Banach space of all bounded families $(x_i) \in \prod_i X_i$ equipped with the norm $\|(x_i)\|_\infty = \sup_i \|x_i\|_{X_i}$. If $\mathcal{U}$ is an ultrafilter on I, then Theorem 2.4 of Chapter 1 will imply that $\lim_{\mathcal{U}} \|x_i\|_{X_i}$ exists. We can then define a seminorm $\mathcal{N}((x_i)) = \lim_{\mathcal{U}} \|x_i\|_{X_i}$ on $\ell_\infty(X_i)$. Ker$\mathcal{N}$, the kernal of $\mathcal{N}$, is given by

$$\ker \mathcal{N} = \{(x_i) \in \ell_\infty(X_i) : \mathcal{N}(x_i) = 0\}. \tag{2.1}$$

Proposition 2.1. Ker$\mathcal{N}$ *is a closed subspace of* $\ell_\infty(X_i)$.

Proof. The properties satisfied by the limit over ultrafilters will guarantee that ker$\mathcal{N}$ is a subspace of $\ell_\infty(X_i)$. To show that it is closed, take a Cauchy sequence $((x_i^n)_{i \in I})_{n \in \mathbb{N}}$ in $\ell_\infty(X_i)$ with $(x_i^n)_{i \in I} \in \ker \mathcal{N}$ for each $n \in \mathbb{N}$. Because $\ell_\infty(X_i)$ is complete, we know that $((x_i^n)_i)_{\mathbb{N}}$ converges to some $(x_i) \in \ell_\infty(X_i)$. Hence let ϵ be an arbitrary positive real number, and consider the nonempty set $J_\epsilon = \{n : \|(x_i^n) - (x_i)\|_\infty \le \epsilon\}$. For any $n \in J_\epsilon$ and $i \in I$ we have $\|x_i^n - x_i\|_{X_i} \le \epsilon$. Then

$$\lim_{\mathcal{U}} \|x_i\|_{X_i} \le \lim_{\mathcal{U}} \|x_i^n\|_{X_i} + \epsilon.$$

Since $(x_i^n) \in \ker \mathcal{N}$, we have $\lim_{\mathcal{U}} \|x_i\|_{X_i} \le \epsilon$, which completes the proof. ∎

Definition 2.2. The *ultraproduct* of the family of Banach ·spaces $(X_i)_{i \in I}$ with respect to the ultrafilter $\mathcal{U}$ on I is the quotient space $\ell_\infty(X_i)/\ker \mathcal{N}$, which will be denoted by $(X_i)_{\mathcal{U}}$. If $X_i = X$ for all $i \in I$, then $(X_i)_{\mathcal{U}} = (X)_{\mathcal{U}}$ is the *ultrapower* of X.

The quotient norm on $(X_i)_{\mathcal{U}}$ is defined as

$$\|\widetilde{(x_i)}\|_{(X_i)_{\mathcal{U}}} = \inf\{\|(x_i + y_i)\|_\infty : (y_i) \in \ker \mathcal{N}\}, \tag{2.2}$$

where $\widetilde{(x_i)}$ is the equivalence class of (x_i). It is a remarkable fact that this norm is independent of the representative (x_i) and can be expressed in a simple form.

Proposition 2.3. *The quotient norm on* $(X_i)_{\mathcal{U}}$ *satisfies*

$$\|\widetilde{(x_i)}\|_{(X_i)_{\mathcal{U}}} = \lim_{\mathcal{U}} \|x_i\|_{X_i} \; \text{for any } \widetilde{(x_i)} \in (X_i)_{\mathcal{U}}. \qquad (2.3)$$

Proof. Let $\widetilde{x} = \widetilde{(x_i)}$ be in $(X_i)_{\mathcal{U}}$; then $\widetilde{x} = \{(x_i + y_i) \in \ell_\infty(X_i) : (y_i) \in \ker\mathcal{N}\}$. Therefore

$$\lim_{\mathcal{U}} \|x_i + y_i\|_{X_i} = \lim_{\mathcal{U}} \|x_i\|_{X_i} \leq \|(x_i + y_i)\|_\infty$$

for any $(y_i) \in \ker\mathcal{N}$, which implies that $\lim_{\mathcal{U}} \|x_i\|_{X_i} \leq \|\widetilde{x}\|_{(X_i)_{\mathcal{U}}}$. Let us prove the other inequality. Consider the set

$$I_\epsilon = \left\{ i \in I : \|x_i\|_{X_i} \leq \lim_{\mathcal{U}} \|x_i\|_{X_i} + \epsilon \right\},$$

where $\epsilon > 0$. By the definition of a limit over $\mathcal{U}$, we have $I_\epsilon \in \mathcal{U}$. Now define (y_i) by setting $y_i = -x_i$ if $i \notin I_\epsilon$ and $y_i = 0$ otherwise. It is easy to verify that $\lim_{\mathcal{U}} \|y_i\|_{X_i} = 0$; so $(x_i + y_i)$ is a representative of $\widetilde{x}$. But $\|(x_i + y_i)\|_\infty = \sup_{i \in I_\epsilon} \|x_i\|_{X_i}$, which implies that

$$\|(x_i + y_i)\|_\infty \leq \lim_{\mathcal{U}} \|x_i\|_{X_i} + \epsilon.$$

Hence $\|\widetilde{x}\|_{(X_i)_{\mathcal{U}}} \leq \|(x_i + y_i)\|_\infty \leq \lim_{\mathcal{U}} \|x_i\|_{X_i} + \epsilon.$

Since ϵ was arbitrary, we obtain the desired inequality. ∎

Remarks 2.4. Proposition 2.3 is crucial to our discussion because it implies that $(X_i)_{\mathcal{U}}$ will inherit any property which can be expressed in terms of norms and which is satisfied in each X_i. For example, if each X_i is a Hilbert space, then $(X_i)_{\mathcal{U}}$ is also a Hilbert space. (Recall that a Banach space X is a Hilbert space if and only if

$$\|x + y\|^2 + \|x - y\|^2 = 2\|x\|^2 + 2\|y\|^2$$

for any $x, y \in X$.) Let us also add that if $\mathcal{U}$ is a trivial ultrafilter generated by $i_0 \in I$, then $(X_i)_{\mathcal{U}}$ is isometrically identical with X_{i_0}.

II. The Banach Space Ultraproduct

If $X_i = X$ for all $i \in I$, then one can embed X isometrically into $(X)_{\mathcal{U}}$. Consider the class of $(x, x, \ldots)$ in $(X)_{\mathcal{U}}$: we have

$$\|(\widetilde{x, x, \ldots})\|_{(X)_{\mathcal{U}}} = \lim_{\mathcal{U}} \|x\| = \|x\|,$$

which means that X can be viewed as a subspace of $(X)_{\mathcal{U}}$. In general, members of $(X)_{\mathcal{U}}$ will be denoted by $\tilde{x}$, but we will use $x \in (X)_{\mathcal{U}}$ to mean $x \in X$ viewed as a subspace of $(X)_{\mathcal{U}}$.

In the case that X is an infinite dimensional space and $\mathcal{U}$ is a nontrivial ultrafilter on the set of natural numbers $\mathbb{N}$, $(X)_{\mathcal{U}}$ contains X as a proper subspace. To show this let (x_n) be a bounded sequence in X with no convergent subsequence, and consider $\widetilde{(x_n)}$ in $(X)_{\mathcal{U}}$. Assume that $\widetilde{(x_n)} \in X$, i.e. that there exists $x \in X$ such that $(x, x, \ldots) = \widetilde{(x_n)}$; this implies that $\lim_{\mathcal{U}} \|x - x_n\| = 0$. Using Remark 2.3 of Chapter 1, one can find a subsequence (x'_n) of (x_n) which converges to x, and this contradicts the hypothesis. Therefore $\widetilde{(x_n)} \notin X$. One can also use this example to show that $(X)_{\mathcal{U}}$ is non-separable, independent of the separability of X.

The next theorem gives an interesting result concerning the existence of a "good" subsequence of a given sequence in a Banach space.

Theorem 2.5. *Suppose that X is a separable Banach space and (x_n) is a bounded sequence which has no convergent subsequence. Then there exists a subsequence (x'_n) of (x_n) such that*

$$\lim_{\substack{n_k < n_{k-1} < \cdots < n_1 \\ n_k \to \infty}} \left\| x + \sum_{i=1}^{k} c_i x'_{n_i} \right\| \tag{2.4}$$

exists for any scalars $(c_i)_{i \leq k}$ and any $x \in X$.

Proof. Let $\mathcal{U}$ be a nontrivial ultrafilter over $\mathbb{N}$ and define a sequence of spaces (X_k) by setting $X_0 = X$ and $X_{k+1} = (X_k)_{\mathcal{U}}$. Since these spaces are nested, the space $Y = \bigcup_k X_k$ is a normed space. The completion of Y will be denoted by X_∞, i.e. $X_\infty = \overline{Y} = \overline{\bigcup_k X_k}$. Since $X \subset X_k$ for any $k \in \mathbb{N}$, the sequence (x_n) can be viewed as a sequence in X_k, so set $\tilde{x}_k = \widetilde{(x_n)}$ in $X_k = (X_{k-1})_{\mathcal{U}}$. Since (x_n)

24

has no convergent subsequence, it follows that $\tilde{x}_k \in X_{k+1} - X_k$. Furthermore, $\tilde{x}_k$ has the property that

$$\left\| x + \sum_{i=1}^{l} a_i \tilde{x}_i \right\|_{X_\infty} = \left\| x + \sum_{i=1}^{l} a_i \tilde{x}_{n_i} \right\|_{X_\infty} \tag{2.5}$$

for any $x \in X$, any finite sequence of scalars $(a_i)_{i \le l}$, and any increasing sequence of integers $(n_i)_{i \le l}$.

To prove (2.5), let us show that

$$\|x + a_1\tilde{x}_1 + a_2\tilde{x}_2 + a_3\tilde{x}_4\|_{X_\infty} = \|x + a_1\tilde{x}_1 + a_2\tilde{x}_3 + a_3\tilde{x}_4\|_{X_\infty}; \tag{2.6}$$

the same argument will work for any finite sum. First, we have

$$\|x + a_1\tilde{x}_1 + a_2\tilde{x}_2 + a_3\tilde{x}_4\|_{X_\infty} = \|x + a_1\tilde{x}_1 + a_2\tilde{x}_2 + a_3\tilde{x}_4\|_{X_4}$$
$$= \lim_{n,\mathcal{U}} \|x + a_1\tilde{x}_1 + a_2\tilde{x}_2 + a_3 x_n\|_{X_3}$$

But for any n

$$\|x + a_3 x_n + a_1\tilde{x}_1 + a_2\tilde{x}_3\|_{X_3} = \lim_{m,\mathcal{U}} \|x + a_3 x_n + a_1\tilde{x}_1 + a_2 x_m\|_{X_2}$$
$$= \lim_{m,\mathcal{U}} \|x + a_3 x_n + a_1\tilde{x}_1 + a_2 x_m\|_{X_1}.$$

On the other hand

$$\|x + a_1\tilde{x}_1 + a_2\tilde{x}_2 + a_3 x_n\|_{X_3} = \|x + a_1\tilde{x}_1 + a_2\tilde{x}_2 + a_3 x_n\|_{X_2}$$
$$= \lim_{m,\mathcal{U}} \|x + a_1\tilde{x}_1 + a_2 x_m + a_3 x_n\|_{X_1},$$

which proves the desired equality (2.6).

To complete the proof of theorem 2.5, we will show that there exists a subsequence (x'_n) of (x_n) such that

$$\left\| x + \sum_{i=1}^{k} a_i \tilde{x}_i \right\|_{X_\infty} = \lim_{\substack{n_k < n_{k-1} < \cdots < n_1 \\ n_k \to \infty}} \left\| x + \sum_{i=1}^{k} a_i x'_{n_i} \right\|. \tag{2.7}$$

But to prove this, it is enough to find, for a given $\epsilon > 0$, a subsequence (x'_n) of (x_n) such that

$$\left\| x + \sum_{i=1}^{k} a_i \tilde{x}_i \right\|_{X_\infty} \underset{\epsilon}{=} \left\| x + \sum_{i=1}^{k} a_i x'_{n_i} \right\|, \tag{2.8}$$

II. The Banach Space Ultraproduct

where "$\underset{\epsilon}{=}$" is defined so that $a \underset{\epsilon}{=} b$ if and only if $|a - b| \leq \epsilon$. We will give the proof for $k = 2$; the general proof uses the same technique.

We have

$$\|x + a_1\tilde{x}_1 + a_2\tilde{x}_2\|_{X_2} = \lim_{n,\mathcal{U}} \|x + a_1\tilde{x}_1 + a_2 x_n\|_{X_1}.$$

Hence there exists a subsequence (x'_{n_k}) of (x_n) such that

$$\|x + a_1\tilde{x}_1 + a_2\tilde{x}_2\|_{X_2} = \lim_{k \to \infty} \|x + a_1\tilde{x}_1 + a_2 x'_{n_k}\|_{X_1},$$

and it follows that there exists $k_0 \geq 0$ such that

$$\|x + a_1\tilde{x}_1 + a_2\tilde{x}_2\|_{X_2} \underset{\epsilon}{=} \|x + a_1\tilde{x}_1 + a_2 x'_{n_k}\|_{X_1}$$

for any $k \geq k_0$. Therefore there exists $I_k \in \mathcal{U}$ for any $k \geq k_0$ such that if $m \in I_k$ then

$$\|x + a_1\tilde{x}_1 + a_2\tilde{x}_2\|_{X_2} \underset{2\epsilon}{=} \|x + a_1 x_m + a_2 x'_{n_k}\|.$$

Since $\mathcal{U}$ is nontrivial, one can then construct a sequence of increasing integers (n_i) satisfying

$$\|x + a_1\tilde{x}_1 + a_2\tilde{x}_2\|_{X_2} \underset{\epsilon}{=} \lim_{\substack{j < i \\ j \to \infty}} \|x + a_1 x_{n_i} + a_2 x_{n_j}\|,$$

which completes the proof of (2.8) when $k = 2$. ∎

Note that the subsequence found in proving (2.7) depends on x, but if X is separable we can find a subsequence which satisfies (2.7) and is independent of x.

Theorem 2.5 leads to the following definition.

Definition 2.6. A Banach space F is called a *spreading model* for X generated by (x_n) provided both of the following hold:

(1) $X \subset F$ and there is a sequence (e_j) such that F is the closed linear span of $X \cup \{e_j\}$. The sequence (e_j) is called the *fundamental basis* of F.

(2) There exists a nonconvergent subsequence (x'_n) of (x_n) such that for any $x \in X$ and any finite scalars $(c_i)_{1 \leq i \leq k}$,

$$\left\| x + \sum_{i=1}^{k} c_i e_i \right\|_F = \lim_{\substack{m_1 < m_2 < \cdots < m_k \\ m_1 \to \infty}} \left\| x + \sum_{i=1}^{k} c_i x'_{n_i} \right\|_X \qquad (2.9)$$

holds. Any nonconvergent sequence (x'_n) for which the limit on the right side of (2.9) exists is called a *good* sequence.

The idea of spreading models, but not the terminology, was first introduced by Brunel and Sucheston [35]. Their construction is based on a concept of extraction using the *Ramsey principle*. Applications of spreading models can be found in [17, 18, 35, 36, 82, 124].

Remarks 2.7. Suppose that X is a separable Banach space and (x_n) is any bounded sequence with no convergent subsequence. Then Theorem 2.5 indicates that (x_n) generates a spreading model. Indeed, if we define $F_0 = X \times \mathbf{R}^{(\mathbf{N})}$ and for any $(x, (a_i)) \in F_0$ we set

$$\left\| (x, (a_i)_{i \leq k}) \right\| = \left\| x + \sum_{i=1}^{k} a_{k-i}\tilde{x}_i \right\|_{X_\infty} \tag{2.10}$$

then the Banach space $F = \overline{F_0}$ (the completion of F_0 with respect to the norm defined by (2.10)) is the spreading model generated by (x_n). Notice that the canonical basis of $\mathbf{R}^{(\mathbf{N})}$ is the fundamental basis (e_i) (see Definition 2.6).

For more details on this construction the reader can consult [18]. Note that by $\mathbf{R}^{(\mathbf{N})}$ we mean the set of those sequences of scalars (a_n) which have only finitely many non-zero terms. We will identify such sequences with finite sequences of scalars.

Let us add that the fundamental basis (e_n) is a Schauder basic sequence (see [18]). Furthermore, (e_n) has the interesting property that

$$\left\| x + \sum_{i=1}^{k} c_i e_i \right\|_F = \left\| x + \sum_{i=1}^{k} c_i e_{n_i} \right\|_F \tag{2.11}$$

for any $x \in X$, finite scalars $(c_i)_{i \leq k}$, and increasing sequence of integers (n_i); using this one can deduce the next result. (For more details see [18], p. 24.)

Proposition 2.8 *Let (x_n) be a spreading bounded sequence in X which is weakly convergent to 0. Then the fundamental sequence (e_n) is an unconditional basis.*

II. The Banach Space Ultraproduct

In the end of this section we will study classes of Banach spaces which are stable under ultraproducts. Let us remark that the classical representation theorems are helpful in proving stability under ultraproducts, but they don't give us any information on the structure of the ultraproduct. The following theorem illustrates this.

Theorem 2.9 *Let I be a set and let $\mathcal{U}$ be an ultrafilter on I.*

(i) Let $(K_i)_{i \in I}$ be compact Hausdorff spaces. Then there is a compact Hausdorff space K such that the ultraproduct $(C(K_i))$ is linearly isometric to $C(K)$. This isometry preserves the multiplicative and lattice structures.

(ii) Let $1 \le p < \infty$ and let $(\mu_i)_{i \in I}$ be a set of arbitrary σ-additive measures. Then $(L^p(\mu_i))_{\mathcal{U}}$ is order-isometric to $L^p(\mu)$ for a certain measure μ.

The proof of statement *(i)* is based on the Gelfand-Naimark representation theorem for commutative C^*-algebras with identity (see [126] for more details), while *(ii)* follows the theorem of Bohnenblust and Nakano [126] which characterizes the L_p-Banach spaces. Of course, for both *(i)* and *(ii)* we use the easy fact that the class of C^*-algebras and the class of Banach lattices are stable under ultraproducts.

In light of Theorem 2.9 one can ask how the compact space K and the measure μ given by statements *(i)* and *(ii)* can be determined. The first half of the question is answered by a result due to Henson [85], which establishes the following:

Theorem 2.10. *Let $(K_i)_{i \in I}$ be a family of compact Hausdorff spaces and let K satisfy $(C(K_i))_{\mathcal{U}} \cong C(K)$ (as in Theorem 2.9); then the set-theoretic ultraproduct $(K_i)_{\mathcal{U}}$ is homeomorphic to a dense subset of K.*

The reader interested in the details of the proof of Theorem 2.10 can consult [84, 174].

In order to complete the answer to the question raised regarding Theorem 2.9, we will carry out some more explicit constructions which will emphasize the relation to the product of measure spaces described in the last section. Let us note that statement *(ii)* of

Theorem 2.9 was first given in [52]. The result described here is due to Heinrich [84], who produced it in the class of Orlicz spaces.

First let $(\Omega_i, \mathcal{A}_i, \mu_i)_{i \in I}$ be a family of σ-additive measure spaces. Without loss of generality we can assume that $(\mu_i)_{i \in I}$ are probabilities. Let $(\widetilde{\Omega}, \widetilde{\mathcal{A}}, \widetilde{\mu})$ be the set-theoretic ultraproduct measure space associated with $(\Omega_i, \mathcal{A}_i, \mu_i)_{i \in I}$, as described in Proposition 1.3. Heinrich [84] proves the following.

Theorem 2.11. *Let $(\Omega_i, \mathcal{A}_i, \mu_i)_{i \in I}$ and $(\widetilde{\Omega}, \widetilde{\mathcal{A}}, \widetilde{\mu})$ be as given above. Then, for $1 \leq p < \infty$,*

$$\left(L^p(\mu_i) \right)_{\mathcal{U}} \cong L^p(\widetilde{\mu}) \oplus_p L^p(\nu),$$

for some measure ν. More specifically, we have:

(i) There is an isometry $J : L^p(\widetilde{\mu}) \to (L^p(\mu_i))_{\mathcal{U}}$.

(ii) There is a norm-one projection $\widetilde{P}$ from $(L^p(\mu_i))_{\mathcal{U}}$ onto the subspace $J(L^p(\widetilde{\mu}))$ which satisfies
(1) $\widetilde{f} \geq \widetilde{P}(\widetilde{f}) \geq 0$ provided that $\widetilde{f} \geq 0$;
(2) $\ker \widetilde{P}$ is a lattice isometric to $L^p(\nu)$ for some measure ν, and

$$\|\widetilde{f}\|^p = \|\widetilde{P}(\widetilde{f})\|^p + \|(I - \widetilde{P})(\widetilde{f})\|^p.$$

Proof. The functions of the form

$$\widetilde{f} = \sum_{k=1}^{n} \alpha_k \chi_{\widetilde{A}_k} \text{ with } \widetilde{A}_k = \widetilde{(A_k^i)} \in \widetilde{\mathcal{A}}_0$$

are dense in $L_p(\widetilde{\Omega}, \widetilde{\Sigma}, \widetilde{\mu})$. We set $J(\widetilde{f}) = (\widetilde{\sum_{k=1}^{n} \alpha_k \chi_{A_k^i}})$. J is an isometry and therefore extends to an isometric embedding of $L^p(\widetilde{\Omega}, \widetilde{\mathcal{A}}, \widetilde{\mu})$ into $(L^p(\Omega_i, \mathcal{A}_i, \mu_i))_{\mathcal{U}}$. We shall identify both spaces in the sequel.

Now let us construct the projection $\widetilde{P}$. Let $\widetilde{f} = \widetilde{(f_i)} \in (L^p(\Omega_i, \mathcal{A}_i, \mu_i))_{\mathcal{U}}$. We define a measure $\widetilde{\nu}_0$ on $\widetilde{\mathcal{A}}_0$ by

$$\widetilde{\nu}_0(\widetilde{A}) = \lim_{\mathcal{U}} \int_{A_i} f_i \, d\mu_i$$

for any $\tilde{A} = \widetilde{(A_i)} \in \tilde{A}_0$. It is proven in [84] that $\tilde{\nu}_0$ possesses a unique extension $\tilde{\nu}$ to all $\tilde{A}$. It is also shown that if $p > 1$ then $\tilde{\nu}$ is $\tilde{\mu}$-continuous, and when $p = 1$ the $\tilde{\mu}$-continuous part of $\tilde{\nu}$ provides us with a function $\tilde{g}$ which is the Radon-Nikodym derivative of $\tilde{\nu}$ with respect to $\tilde{\mu}$. One can put $\tilde{P}(\tilde{f}) = \tilde{g}$ and derive all the properties of $\tilde{P}$ stated in Theorem 2.11. For more details, the reader can consult [84]. ∎

Let us point out a detail of the projection $\tilde{P}$ which will be of interest in the last chapter. We restrict our attention to the case $p = 1$, and consider the spaces $(L^1(\Omega, \mathcal{A}, \mu))_{\mathcal{U}}$ and $L^1(\tilde{\Omega}, \tilde{\mathcal{A}}, \tilde{\mu})$, where $\mathcal{U}$ is a nontrivial ultrafilter on $\mathbb{N}$. We will need the following definition.

Definition 2.12. Let F be a subset of $L^1(\Omega, \mathcal{A}, \mu)$. We shall say that the functions in F are *equi-integrable* if

$$\sup_{f \in F} \int_{\{|f(\omega)| > a\}} |f(\omega)| \, d\mu(\omega) \longrightarrow 0 \text{ as } a \to \infty.$$

A basic property of equi-integrability is given in the following theorem.

Theorem 2.13 *A subset F of L^1 is equi-integrable if and only if F is relatively compact under the weak topology $\sigma(L^1, L^\infty)$.*

More on equi-integrability is given in [16, p. 153].

Let (f_n) be an equi-integrable sequence in $L^1(\mu)$. Consider $\tilde{f} = \widetilde{(f_n)}$ in $(L^1(\mu))_{\mathcal{U}}$ – one can easily prove that $\|\tilde{P}\tilde{f}\| = \lim_{\mathcal{U}} \|f_n\| = \|\tilde{f}\|$. Therefore $\|(I - \tilde{P})\tilde{f}\| = 0$, which implies that $\tilde{f} \in L^1(\tilde{\mu})$.

This yields the following result.

Proposition 2.14 *Let F be an equi-integrable subset of $L^1(\mu)$. Then $\tilde{F}$ is a subset of $L^1(\tilde{\mu})$, where*

$$\tilde{F} = \{\tilde{f} \in (L^1(\mu))_{\mathcal{U}} : \text{there exists a representative } (f_n)$$
$$\text{of } \tilde{f} \text{ such that } f_n \in F \text{ for all } n\}.$$

III. FINITE REPRESENTABILITY

The concept of finite representability was introduced by James ([94]). It found many applications in the local theory. We will start with a couple of definitions.

Definition 3.1. Let X and Y be Banach spaces and let $0 < \epsilon < 1$. We say that $T : X \to Y$ is an *ϵ-isometry* if

$$(1 - \epsilon)\|x\| \leq \|Tx\| \leq (1 + \epsilon)\|x\|$$

holds for all $x \in X$.

Definition 3.2. Again let X and Y be Banach spaces; we say that X is *finitely representable* in Y if and only if for every $0 < \epsilon < 1$ and every finite dimensional subspace X_0 of X, there exists a finite dimensional subspace Y_0 of Y such that

a) $\dim X_0 = \dim Y_0$;

b) there exists an ϵ-isometry from X_0 to Y_0.

If there exists an ϵ-isometry from X to Y then there exists an isomorphic mapping $T : X \to Y$ such that $\|T\| \, \|T^{-1}\| \leq \frac{1+\epsilon}{1-\epsilon}$, and the converse is also true. This fact can be expressed in terms of the Banach-Mazur distance.

Definition 3.3. Let X and Y be Banach spaces. The *Banach-Mazur distance* between X and Y is denoted by $d(X, Y)$ and is defined as

$$d(X, Y) = \inf\{\|T\| \, \|T^{-1}\| : T \text{ is an isomorphism from } X \text{ onto } Y\}.$$
$$(3.1)$$

When X and Y are not isomorphic we simply set $d(X, Y) = \infty$.

Therefore there exists an ϵ-isometry from X onto Y if and only if $d(X, Y) < \frac{1+\epsilon}{1-\epsilon}$. Hence X is finitely representable in Y if for any $0 < \eta < 1$ and for any finite dimensional subspace X_0 of X, there exists a finite dimensional subspace Y_0 of Y such that $\dim X_0 = \dim Y_0$ and $d(X_0, Y_0) \leq 1 + \eta$.

In general it is not easy to tell whether a Banach space X is finitely representable in a Banach space Y. However, Dvoretzky [59] has proven that ℓ_2 is finitely representable in any Banach space.

In the next theorems a connection between finite representability and ultrapowers is given.

Theorem 3.4. *Let $(X_k)_{k \in I}$ be a family of Banach spaces and $\mathcal{U}$ a nontrivial ultrafilter on I. Let $\widetilde{M}$ be a finite dimensional subspace of $(X_k)_{\mathcal{U}}$. Then for any $0 < \epsilon < 1$, there exists $I_\epsilon \in \mathcal{U}$ such that for each $k \in I_\epsilon$, one can find a finite dimensional subspace M_k of X_k with*

$$d(\widetilde{M}, M_k) \leq \frac{1 + \epsilon}{1 - \epsilon}.$$

Proof. Let $(\widetilde{x}_i)_{1 \leq i \leq n}$ be the unit basis vectors of $\widetilde{M}$. Choose representatives $(x_i^k)_{k \in I}$ of the $\widetilde{x}_i$ such that

$$\|x_i^k\|_{X_k} \leq 2 \tag{3.2}$$

for each k. Let M_k be the finite dimensional subspace of X_k defined by $M_k = \operatorname*{span}_{1 \leq i \leq n} (x_i^k)$. Now define $T_k : \widetilde{M} \to M_k$ by

$$T_k\left(\sum_{i=1}^n a_i \widetilde{x}_i\right) = \sum_{i=1}^n a_i x_i^k$$

for every $k \in I$. Set $\alpha = \sup\{\sum |a_i| : \| \sum_{i=1}^n a_i \widetilde{x}_i\|_{(X_k)_{\mathcal{U}}} \leq 1\}$. Using (3.2) one can easily verify that $\|T_k\| \leq 2\alpha$. Suppose as before that $0 < \epsilon < 1$, and associate to $\widetilde{x} \in \widetilde{M}$ the set $I_{\widetilde{x}} \in \mathcal{U}$ given by

$$I_{\widetilde{x}} = \left\{ k \in I : (1 - \frac{\epsilon}{2})\|\widetilde{x}\| \leq \|T_k(\widetilde{x})\|_{X_k} \leq \|\widetilde{x}\| \right\}. \tag{3.3}$$

Put $\delta = \frac{\epsilon}{2\alpha}$. $\widetilde{M}$ is a finite dimensional subspace of $(X_k)_{\mathcal{U}}$, so there exists a finite δ-net $(\widetilde{y}_j)_{j \leq m}$ in the unit sphere of $\widetilde{M}$, i.e. for any $\widetilde{x} \in \widetilde{M}$ with $\|\widetilde{x}\| = 1$, there exists $1 \leq j \leq m$ such that $\|\widetilde{x} - \widetilde{y}_j\| \leq \delta$.

On the other hand, $I_0 = \bigcap\limits_{1 \le j \le m} I_{\widetilde{y}_j} \in \mathcal{U}$ and for any $k \in I_0$ and $\widetilde{x} \in \widetilde{M}$ with $\|\widetilde{x}\| = 1$ we have

$$\|T_k \widetilde{y}_j\| - \|T_k(\widetilde{x} - \widetilde{y}_j)\| \le \|T_k \widetilde{x}\| \le \|T_k(\widetilde{x} - \widetilde{y}_j)\| + \|T_k \widetilde{y}_j\| \quad (3.4)$$

for all $1 \le j \le m$. Since $(\widetilde{y}_j)_{j \le m}$ is a δ-net and $k \in I_0$, we obtain

$$1 - \frac{\epsilon}{2} - 2\alpha\delta \le \|T_k \widetilde{x}\| \le 1 + \frac{\epsilon}{2} + 2\alpha\delta.$$

This implies that T_k is one-to-one and $d(\widetilde{M}, M_k) \le \|T_k\| \|T_k^{-1}\|$ $\le \frac{1+\epsilon}{1-\epsilon}$. ∎

Theorem 3.4 gives us a wide class of Banach spaces which are finitely representable in X. The next theorem completes this result by showing that spaces which are finitely representable in X are exactly the subspaces of ultrapowers of X.

Theorem 3.5. *Let X and Y be Banach spaces and suppose that Y is finitely representable in X. Then there is an ultrafilter $\mathcal{U}$ on a set I such that Y is isometrically isomorphic to a subspace of $(X)_{\mathcal{U}}$.*

Proof. Assume Y is not finite dimensional. (When Y is a finite dimensional space, it is easy to show that Y is isometric to a subspace of $(X)_{\mathcal{U}}$, where $\mathcal{U}$ is any nontrivial ultrafilter on $\mathbf{N}$.) Let I be the set of pairs (M, ϵ), where M is a finite dimensional subspace of Y and $0 < \epsilon < 1$. Order I by setting

$$(M, \epsilon) \prec (M', \epsilon') \text{ if and only if } M \subset M' \text{ and } \epsilon' \le \epsilon.$$

It is not hard to verify that $(I, \prec)$ is a lattice, i.e. any two elements in I have a greatest lower bound and a least upper bound in I. Let

$$\mathcal{B} = \big\{ B(i_0) \subset I : B(i_0) = \{ i \in I : i_0 \prec i \} \big\}.$$

Using the lattice structure of I, it is easily seen that $\mathcal{B}$ has the finite intersection property. Consider the filter $\mathcal{F}(\mathcal{B})$ defined in Proposition 1.3 of Chapter 1, namely

$$\mathcal{F}(\mathcal{B}) = \{ S \in 2^I : \exists i_0 \in I \text{ such that } B(i_0) \subset S \}.$$

Since Y is not finite dimensional, $\phi \notin \mathcal{F}(\mathcal{B})$, hence let $\mathcal{U}$ be an ultrafilter containing $\mathcal{F}(\mathcal{B})$. Also, since Y is finitely representable in X, for any $i = (M_i, \epsilon_i) \in I$ there exists an ϵ_i-isometry T_i from M_i onto a finite dimensional subspace X_i of X.

Consider the mapping $J : Y \to (X_i)_\mathcal{U}$ defined by $J(y) = \widetilde{(y_i)}$, where $y_i = T_i(y)$ if $y \in M_i$ and $y_i = 0$ otherwise. The mapping J is a linear isometry. Indeed, let y, y' be in Y and fix $0 < \epsilon_0 < 1$. Set $i_y = ($ span$(y), \epsilon_0)$ and $i_{y'} = ($ span$(y'), \epsilon_0)$. Then $I_{y,y'} = B(i_y) \cap B(i_{y'}) \in \mathcal{U}$. For any $i \in I_{y,y'}$, we have $(J(y))_i = T_i(y)$ and $(J(y'))_i = T_i(y')$. The linearity of T_i implies that

$$\left(J(\alpha y + \beta y')\right)_i = \alpha \left(J(y)\right)_i + \beta \left(J(y')\right)_i$$

for any scalars α, β. And since $I_{y,y'} \in \mathcal{U}$ it follows that

$$J(\alpha y + \beta y') = \alpha J(y) + \beta J(y').$$

To complete the proof we must show that J is an isometry. Let $0 < \epsilon < 1$ and pick $y \in Y$. Put $i_0 = ($span$(y), \epsilon)$. Then $B(i_0) \in \mathcal{U}$ and for any $i = (M_i, \epsilon_i) \in B(i_0)$ we have

$$(1 - \epsilon_i)\|y\|_Y \leq \|T_i(y)\|_X \leq (1 + \epsilon_i)\|y\|_Y,$$

but since $\epsilon_i \leq \epsilon$ this becomes

$$(1 - \epsilon)\|y\|_Y \leq \|T_i(y)\|_X = \|(J(y))_i\|_X \leq (1 + \epsilon)\|y\|_Y.$$

This is true for any i; therefore,

$$(1 - \epsilon)\|y\|_Y \leq \lim_\mathcal{U} \|(J(y))_i\|_X \leq (1 + \epsilon)\|y\|_Y.$$

This implies that J is an ϵ-isometry, for any $0 < \epsilon < 1$. Hence the proof is complete. ∎

Finite representability is clearly a transitive relation, so any subspace of an ultrapower of X is finitely representable in X.

Using the same arguments as in the proof of Theorem 3.5, one can also obtain the following

Proposition 3.6 *Any spreading model of a Banach space X is finitely representable in X.*

IV. SUPER-(M)-PROPERTIES AND BANACH-SAKS PROPERTIES

We now turn to properties of a Banach space which are determined by its finite-dimensional subspaces.

Definition 4.1. Let "$\mathcal{P}$" be a property defined on a Banach space X. We say that X has the property "*super-$\mathcal{P}$*" (resp. "*M-$\mathcal{P}$*") if every finitely representable Banach space in X has property "$\mathcal{P}$" (resp. any spreading model of X has "$\mathcal{P}$").

For instance, a Banach space X is super-reflexive if and only if every finitely representable Banach space in X is reflexive.

Let us remark that in case the property "$\mathcal{P}$" is hereditary, i.e. any subspace of a space that has "$\mathcal{P}$" must also have "$\mathcal{P}$", one can use Theorem 3.5 of the previous section to prove that X has super-P if every ultrapower of X has $\mathcal{P}$.

In general it is difficult to determine whether a given property $\mathcal{P}$ is a superproperty, or to find the associated property super-$\mathcal{P}$. In order to give some examples, we introduce a property which was shown by Clarkson [50] to hold in all the standard ℓ_p and L_p spaces for $1 < p < \infty$.

Definition 4.2. A Banach space X is said to be *uniformly convex* if for each $\epsilon > 0$ there exists $\delta(\epsilon) > 0$ such that for any $x, y \in X$ the conditions $\|x\| \leq 1$, $\|y\| \leq 1$, $\|x - y\| \geq \epsilon$ imply

$$\frac{1}{2}\|x + y\| \leq 1 - \delta(\epsilon). \tag{4.1}$$

Obviously, uniformly convex spaces are strictly convex. (X is said to be strictly convex if $\|x\| \leq 1$, $\|y\| \leq 1$, $\|x - y\| > 0$ imply that $\frac{1}{2}\|x + y\| < 1$ for any $x, y \in X$.)

More generally, X is strictly convex if and only if $\delta(2) = 1$ satisfies Definition 4.2.

The following will be useful in studying these geometric properties more systematically.

IV. Super-(M)-Properties and Banach-Saks Properties

Definition 4.3. The *modulus of convexity* of a Banach space X is the function $\delta_X : [0,2] \to [0,1]$ defined by

$$\delta_X(\epsilon) = \inf\left\{1 - \left\|\frac{x+y}{2}\right\| : \|x\| \le 1, \|y\| \le 1, \|x-y\| \ge \epsilon\right\}. \quad (4.2)$$

The *characteristic of uniform convexity* ϵ_0 of X is defined by

$$\epsilon_0(X) = \sup\{\epsilon : \delta_X(\epsilon) = 0\}.$$

So a Banach space X is uniformly convex if and only if $\delta_X(\epsilon) > 0$ for all $\epsilon > 0$. In the next theorem we consider the modulus of convexity of an ultrapower of a Banach space.

Theorem 4.4. *Let X be a Banach space and $\mathcal{U}$ an ultrafilter on $\mathbf{N}$. Then for any $\epsilon > 0$ we have*

$$\delta_X(\epsilon) = \delta_{(X)_{\mathcal{U}}}(\epsilon). \quad (4.3)$$

Proof. Let $\epsilon > 0$ be fixed. Since $(X)_{\mathcal{U}}$ contains X as a subspace, it follows easily that

$$\delta_{(X)_{\mathcal{U}}}(\epsilon) \le \delta_X(\epsilon).$$

In order to prove the equality (4.3), let $\tilde{x}$ and $\tilde{y}$ be in $(X)_{\mathcal{U}}$ such that $\|\tilde{x}\| \le 1$, $\|\tilde{y}\| \le 1$, $\|\tilde{x} - \tilde{y}\| \ge \epsilon$. Fix $0 < t < 1$; by using the definition of a limit over an ultrafilter one can find representatives (x_n) and (y_n) of $\tilde{x}$ and $\tilde{y}$ and a subset $I \in \mathcal{U}$ such that for any $n \in I$, $\|x_n\| \le 1$, $\|y_n\| \le 1$, $\|x_n - y_n\| \ge t\epsilon$. Hence, by the definition of $\delta_X(\epsilon)$, we obtain

$$\frac{1}{2}\|x_n + y_n\| \le 1 - \delta_X(t\epsilon)$$

for all $n \in I$. Therefore

$$\frac{1}{2}\|\tilde{x} + \tilde{y}\| \le 1 - \delta_X(t\epsilon)$$

holds. In other words, we have shown that

$$\delta_X(t\epsilon) \le \delta_{(X)_{\mathcal{U}}}(\epsilon)$$

for any $0 < t < 1$. Using the continuity of δ (see [143], vol II.), one can deduce that

$$\delta_X(\epsilon) \leq \delta_{(X)_\mathcal{U}}(\epsilon).$$

This completes the proof. ∎

From Theorem 4.4, we deduce the next result, which gives us an example of a property and its associated superproperty.

Theorem 4.5. *Let X be a Banach space and $\mathcal{U}$ a nontrivial ultrafilter on $\mathbf{N}$. Then the following statements are equivalent:*

(i) $(X)_\mathcal{U}$ is strictly convex;

(ii) $(X)_\mathcal{U}$ is uniformly convex;

(iii) X is uniformly convex.

Proof. The implications *(ii)* $\Rightarrow$ *(i)* and *(ii)* $\Rightarrow$ *(iii)* are evident. On the other hand, Theorem 4.4 establishes that *(iii)* $\Rightarrow$ *(ii)*. To complete the proof, let us show that *(i)* $\Rightarrow$ *(iii)*.

Assume to the contrary that X is not uniformly convex. Then there exist an $\epsilon > 0$ and sequences (x_n), (y_n) contained in the unit ball of X, such that $\|x_n + y_n\| \to 2$ as $n \to \infty$ and $\|x_n - y_n\| \geq \epsilon$. Let $\widetilde{x} = \widetilde{(x_n)}$ and $\widetilde{y} = \widetilde{(y_n)}$ be in $(X)_\mathcal{U}$. Then we have

$$\|\widetilde{x}\| \leq 1, \|\widetilde{y}\| \leq 1, \|\widetilde{x} - \widetilde{y}\| \geq \epsilon, \text{ and } \|\widetilde{x} + \widetilde{y}\| = 2.$$

This contradicts the strict convexity of $(X)_\mathcal{U}$. Therefore, the proof of Theorem 4.5 is complete. ∎

We easily obtain from Theorem 4.5 that super-strict convexity is the uniform convexity property. We now discuss a property which was introduced by James [90].

Definition 4.6. A Banach Space X is said to be *uniformly non-square* if there exists a positive δ such that, for all x, y in the unit ball, the conditions $\|\frac{x+y}{2}\| \geq 1 - \delta$ and $\|\frac{x-y}{2}\| \geq 1 - \delta$ are not simultaneously true, that is,

$$\text{if } \|\frac{x-y}{2}\| \geq 1 - \delta \text{ then } \|\frac{x+y}{2}\| \leq 1 - \delta.$$

IV. Super-(M)-Properties and Banach-Saks Properties

It is easy to verify that a Banach space X is uniformly non-square if $\epsilon_0(X) < 2$. Therefore uniform non-squareness is a super-property. James has also proven that a uniformly non-square Banach space is reflexive, and hence super-reflexive. More on this topic will be given in the "Notes on Normal Structure" section of Chapter 3.

One can easily deduce from Proposition 3.6 that if a Banach space has super-$\mathcal{P}$ then it has M-$\mathcal{P}$, but in general these properties are not equivalent. In order to see this, consider

$$X = \oplus_2 \ell_1^n = \big\{ (x_n) : x_n \in \ell_1^n \text{ and } \sum \|x_n\|^2 < \infty \big\}.$$

X cannot be super-reflexive, since any ultrapower $(X)_\mathcal{U}$ of X, where $\mathcal{U}$ is a non-trivial ultrafilter on $\mathbf{N}$, contains ℓ_1 isometrically. However, X is M-reflexive; in [167] it is shown that any spreading model of X is isometric to $X \oplus_2 \ell_2$, and since X is obviously reflexive, one can then deduce that any spreading model of X is reflexive.

We now introduce the notion of Banach-Saks property, which will be needed in Chapter 3.

Definition 4.7. A Banach space X is said to have the *Banach-Saks property* (BSP) (resp. the *alternate Banach-Saks property* (ABSP)) if for any bounded sequence (x_n) one can find a subsequence (x_n') such that $(\frac{1}{n} \sum_{k=1}^{n} x_k')$ is convergent (resp. $(\frac{1}{n} \sum_{k=1}^{n} (-1)^k x_k')$ is convergent).

The Banach-Saks property was introduced by Banach and Saks [13], who showed that L_p, $1 < p < \infty$, has this property. Later Brunel and Sucheston introduced the alternate Banach-Saks property [35]. The motivation for studying these properties is to be able to guarantee the strong convergence of an approximate solution of a given equation. Indeed, if (x_n) is a sequence of approximate solutions, then it very often happens that a weak cluster point x of (x_n) is an exact solution; hence there exists a subsequence (x_n') which is weakly convergent to x. But weak convergence is not useful for any kind of computational algorithm, whereas if the space has BSP then x can be approximated by the Cesaro sum $(\frac{1}{n} \sum_{k=1}^{n} x_k')$, since in this case the convergence is strong. The connection of these properties to the geometry of Banach spaces can be found in [17, 18, 165].

Now let us turn our attention to the relationship between spreading models of a given Banach space X and the Banach-Saks properties. The next result, which characterizes Banach spaces with the alternate Banach-Saks property, is due to Beauzamy [17].

Theorem 4.8. *A Banach space has ABSP if and only if it does not have a spreading model isomorphic to ℓ_1.*

We omit the proof. The reader can consult [18] for details. The next result, of Guerre and Lapreste, will also be given without proof.

Proposition 4.9. *Let (x_n) be a spreading sequence in X. Consider the fundamental sequence (e_n) of the spreading model F generated by (x_n). If (e_n) is not equivalent to the basis of ℓ_1, then (e_n) converges weakly in F if and only if (x_n) converges weakly in X, and the weak limits are equal.*

V. THE ULTRAPRODUCT OF MAPPINGS

Let $(X_i)_{i \in I}$ and $(Y_i)_{i \in I}$ be two families of Banach spaces indexed by a set I, let $\mathcal{U}$ be an ultrafilter on I, and let $(X_i)_{\mathcal{U}}$ and $(Y_i)_{\mathcal{U}}$ be the ultraproducts of the given families. Consider a family of mappings (T_i) where for each $i \in I$,

$$T_i : D_i \subset X_i \to Y_i.$$

From the family of subsets $(D_i)_{i \in I}$ one can generate a subset $\widetilde{D} = (D_i)_{\mathcal{U}}$ of $(X_i)_{\mathcal{U}}$ defined by

$$\widetilde{D} = \{\widetilde{d} \in (X_i)_{\mathcal{U}} : \exists \text{ a representative } (d_i) \text{ of } \widetilde{d} \\ \text{with } d_i \in D_i \text{ for each } i \in I\}. \tag{5.1}$$

Proposition 5.1 gives some properties which $\widetilde{D}$ may inherit from the family (D_i).

Proposition 5.1. *The following are true:*
(i) $\widetilde{D}$ is convex if the D_i's are all convex;

V. The Ultraproduct of Mappings

(ii) $\tilde{D}$ is closed if the D_i's are all closed;
(iii) $\tilde{D}$ is bounded if the D_i's are all bounded, and $\operatorname{diam}\tilde{D} = \lim_{\mathcal{U}} \operatorname{diam}D_i$.

Definition 5.2. Let $(T_i)_{i \in I}$ be a family of mappings defined on $(D_i)_{i \in I}$. The *ultraproduct mapping* $\tilde{T} = (T_i)_{\mathcal{U}} : (X_i)_{\mathcal{U}} \to (Y_i)_{\mathcal{U}}$ of the mappings $(T_i)_{i \in I}$ is defined on $\tilde{D} = (D_i)_{\mathcal{U}}$ by

$$\tilde{T}(\tilde{d}) = (\widetilde{T_i(d_i)}) \text{ for any } \tilde{d} \in \tilde{D}, \tag{5.2}$$

provided that the equality (5.2) is well-defined, i.e. the family (T_i) satisfies the following condition:

For every $(d_i), (d_i') \in (D_i)$, if $\lim_{\mathcal{U}} \|d_i - d_i'\| = 0$,

$$\text{then } \lim_{\mathcal{U}} \|T_i d_i - T_i d_i'\| = 0. \tag{5.3}$$

When $T_i = T$ for all $i \in I$, $\tilde{T}$ is called the *ultrapower mapping* of T.

Notice that condition (5.3) implies a kind of equicontinuity of the family $(T_i)_{i \in I}$.

Let us discuss an important example which will be of interest in the next chapter. Recall that a mapping $T : D \to Y$ is said to be *Lipschitzian* with *Lipschitz constant* λ if

$$\|T(d) - T(d')\| \leq \lambda \|d - d'\|$$

holds for all $d, d' \in D$.

Let $(T_i)_{i \in I}$ be a family of Lipschitzian mappings with Lipschitz constant $(\lambda_i)_{i \in I}$. Assume that $\lambda = \lim_i \lambda_i$ is finite. Then one can easily show that $(T_i)_{i \in I}$ satisfies condition (5.3). Furthermore, the ultraproduct mapping $\tilde{T}$ of $(T_i)_{i \in I}$ is Lipschitzian with Lipschitz constant λ.

The next result reveals a structural stability under ultraproducts when the mappings involved are linear.

Proposition 5.3. *Assume that $(T_i)_{i \in I}$ are bounded linear operators with* $\sup_{i \in I} \|T_i\| < \infty$. *Then* $\tilde{T} = (T_i)_{\mathcal{U}}$ *is a bounded linear operator taking $(X_i)_{\mathcal{U}}$ into $(Y_i)_{\mathcal{U}}$, with*

$$\|\tilde{T}\| = \lim_{\mathcal{U}} \|T_i\|. \tag{5.4}$$

Proof. The fact that $\widetilde{T}$ is a bounded linear operator with $\|\widetilde{T}\| \leq \lim_{\mathcal{U}} \|T_i\|$ is clear. To show that $\lim_{\mathcal{U}} \|T_i\| \leq \|\widetilde{T}\|$, pick $\epsilon > 0$ and for each $i \in I$ find a unit vector $x_i \in X_i$ such that $(1-\epsilon)\|T_i\| \leq \|T_i(x_i)\|$. Let $\widetilde{x} = \widetilde{(x_i)} \in (X_i)_{\mathcal{U}}$. Then $\|\widetilde{x}\| = 1$ and

$$(1 - \epsilon)\lim_{\mathcal{U}} \|T_i\| \leq \lim_{\mathcal{U}} \|T_i x_i\| = \|\widetilde{T}\widetilde{x}\|.$$

So we have $(1 - \epsilon)\lim_{\mathcal{U}} \|T_i\| \leq \|\widetilde{T}\|$, and since ϵ was arbitrary the conclusion follows. ∎

Let us give an application of Proposition 5.3 which will bring out a relationship between the ultraproduct and duality. Let X be a Banach space and let X^* be its dual. Consider a family $(x_i^*)_{i \in I}$ in X^* with bounded norms. Then by Proposition 5.3, $\widetilde{x}^* = \widetilde{(x_i^*)}$ defines a linear functional which is bounded, i.e. $\widetilde{x}^* \in (X)_{\mathcal{U}}^*$. One can then ask whether all the elements of $(X)_{\mathcal{U}}^*$ can be obtained in this way; in other words, is it true that

$$(X^*)_{\mathcal{U}} = (X)_{\mathcal{U}}^*. \tag{5.5}$$

Unfortunately, the answer is no. Indeed, let X be a reflexive Banach space which is not super-reflexive. Then there exists a nonreflexive Banach space which is finitely representable in X. Therefore there is an ultrapower $(X)_{\mathcal{U}}$ of X which is not reflexive, so one cannot expect a relation like $(X^*)_{\mathcal{U}} = (X)_{\mathcal{U}}^*$ to hold in general.

The next result characterizes when equation (5.5) holds. Let us recall that an ultrafilter $\mathcal{U}$ on a set I is said to be countably incomplete if and only if there exists a sequence (I_n) of elements of $\mathcal{U}$ with $\bigcap_n I_n \notin \mathcal{U}$. (It was shown in Chapter I, Proposition 1.9, that any nontrivial ultrafilter on $\mathbb{N}$ is countably incomplete.)

Theorem 5.4. *Let $(X_i)_{i \in I}$ be a family of Banach spaces and let $\mathcal{U}$ be a countably incomplete ultrafilter on I. Then $(X_i^*)_{\mathcal{U}} \cong (X_i)_{\mathcal{U}}^*$ if and only if $(X_i)_{\mathcal{U}}$ is reflexive.*

The proof is omitted but can be found in [174]. As a direct corollary we obtain

<u>*Corollary 5.5*</u> If $\mathcal{U}$ is a countably incomplete ultrafilter, then $(X)_{\mathcal{U}}^* \cong$ $(X^*)_{\mathcal{U}}$ if and only if X is super-reflexive.

VI. TZIRELSON AND JAMES BANACH SPACES

When one tries to find the spreading model, for example, of a classical Banach space such as ℓ_p or c_0, the search is simplified by the fact that the canonical basis is itself spreading. One can expect some difficulties when dealing with non-classical Banach spaces. We will now introduce two non-classical Banach spaces which play important roles in finding negative answers to a number of conjectures about Banach spaces.

We start by introducing the Tzirelson space [195]. Let $\{A_i : 1 \leq i \leq k\}$ be a family of finite, consecutive subsets of $\mathbb{N}-\{0\}$. We say that this family is *admissible* if $k \leq \min(A_1)$. Note that this condition does not restrict the structures of the A_i's. Each can be arbitrarily long, and there may be gaps between their members. Let $(x_n) \in \mathbb{R}^{(\mathbb{N})}$ and set

$$\begin{cases} \|(x_k)\|_0 = \max_k |x_k|, \text{ and} \\ \|(x_k)\|_{n+1} = \max\{\|(x_k)\|_n : \max_A \tfrac{1}{2}\sum_j \|P_j x\|_n\}, \end{cases} \qquad (6.1)$$

where max means that the maximum is taken over all admissible subsets (A_i) of $\mathbb{N}$. P_j is the natural projection over A_j, i.e.

$$P_j(x) = P_j\left(\sum_k x_k e_k\right) = \sum_{k \in A_j} x_k e_k,$$

where (e_k) is the canonical basis of $\mathbb{R}^{(\mathbb{N})}$. (Recall the definition of $\mathbb{R}^{(\mathbb{N})}$ from Remark 2.7 of section II.)

By induction, one can easily prove that $(\|x\|_n)$ exists, is increasing, and is bounded above by $\|x\|_{\ell_1} = \sum_k |x_k|$.

We set
$$\|x\|_\tau = \lim_n \|x\|_n \text{ for any } x \in \mathbb{R}^{(\mathbb{N})}. \qquad (6.2)$$

Definition 6.1. The Tzirelson space $\mathcal{T}$ is the completion of $\mathbb{R}^{(\mathbb{N})}$ with respect to the norm $\|\cdot\|_\tau$ defined by (6.2).

Observe that if we let $n \to \infty$ in (6.1), we obtain

$$\|x\|_T = \max\{\|x\|_{c_0}, \max_A \{\frac{1}{2}\sum_j \|P_j(x)\|_T\}\} \qquad (6.3)$$

for any $x \in T$, where $\|x\|_{c_0} = \max_k |x_k|$.

The structure of the Tzirelson space T is well understood. For more on this space, the reader may consult the recent book of Cassazza and Shura [46] as well as [18, 40, 42, 43, 45, 67]. We will only emphasize the properties of T which will be of use to us later.

First of all, it is clear from (6.1) that the canonical basis of $\mathbf{R}^{(\mathbf{N})}$ is an unconditional Schauder basis of T and the unconditional constant is 1.

Next we will prove that T is reflexive. Let $(u_i)_{1 \leq i \leq 2n}$ be a finite sequence of unit vectors in T which form consecutive blocks of the canonical basis. Let B_j denote the support of u_j for $j = n+1$, $n+2, \ldots, 2n$. Hence $(B_j)_{n+1 \leq j \leq 2n}$ forms a sequence of admissible subsets of $\mathbf{N}$. Using (6.3) we then obtain

$$\| \sum_{i=n+1}^{n+n} a_i u_i \|_T \geq \frac{1}{2}\sum_j \| P_{B_j}(\sum_{i=n+1}^{2n} a_i u_i) \|_T$$

$$= \frac{1}{2}\sum_j \| P_{B_j}(a_j u_j) \|_T$$

which implies

$$\frac{1}{2}\sum_{j=n+1}^{2n} |a_j| \leq \| \sum_{i=n+1}^{2n} a_i u_i \|_T \leq \sum_{i=n+1}^{2n} |a_i|. \qquad (6.4)$$

So the sequence (u_j) is equivalent to the canonical basis of ℓ_1^n. Therefore T cannot contain c_0 or ℓ^p for $p > 1$. Indeed, if it did, there would exist consecutive blocks in T which would be equivalent to the basis of c_0 or ℓ^p, by Proposition 0.14 of Chapter 0; this would contradict (6.4).

To prove that T does not contain ℓ_1, we first need the following proposition.

Proposition 6.2. *Let $r \geq 2$ be an integer and let $u_0, u_1, \ldots, u_r$ be a sequence of consecutive blocks in T with $\|u_i\| = 1$ for all i. Then*

$$\left\| u_0 + \frac{1}{r} \sum_{i=1}^{r} u_i \right\|_T \leq \frac{7}{4}, \tag{6.5}$$

provided that $\max B_0 \leq \frac{r}{2}$, where B_0 is the support of u_0.

Proof. Clearly, we have $\left\| u_0 + \frac{1}{r} \sum_{i=1}^{r} u_i \right\|_{co} \leq 1$. Let $(A_j)_{1 \leq j \leq k}$ be a family of admissible subsets of $\mathbf{N}$. First we assume that $\min A_1 > \max B_0$. Then

$$\frac{1}{2} \sum_{j=1}^{k} \left\| P_j(u_0 + \frac{1}{r} \sum_{i=1}^{r} u_i) \right\|_T = \frac{1}{2} \sum_{j=1}^{k} \left\| P_j(\frac{1}{r} \sum_{i=1}^{r} u_i) \right\|_T,$$

which implies that $\frac{1}{2} \sum_{j=1}^{k} \left\| P_j(u_0 + \frac{1}{r} \sum_{i=1}^{r} u_i) \right\|_T \leq 1$.

Now let us assume that $\min A_1 < \max B_0$. Since (A_j) is admissible and $\max B_0 \leq \frac{r}{2}$, we have $k \leq \frac{r}{2}$. Define

$$\Delta = \left\{ i \geq 1 : \|P_j u_i\|_T \neq 0 \text{ for at least two indices } j \right\}$$

and

$$\Gamma = \left\{ i \geq 1 : \|P_j u_i\|_T \neq 0 \text{ for at least one index } j \right\}.$$

Then clearly $\operatorname{card}(\Delta) \leq \frac{r}{2} - 1$, and

$$\frac{1}{2} \sum_{j} \left\| P_j(u_0 + \frac{1}{r} \sum_{i=1}^{r} u_i) \right\|_T \leq \frac{1}{2} \sum_{j} \left\| P_j(u_0) \right\|_T +$$
$$\frac{1}{2r} \sum_{i \in \Delta} \sum_{j} \left\| P_j(u_j) \right\|_T + \frac{1}{2r} \sum_{i \in \Gamma} \sum_{J} \left\| P_j(u_j) \right\|_T.$$

Therefore,

$$\frac{1}{2} \sum_{j} \left\| P_j(u_0 + \frac{1}{r} \sum_{i=1}^{r} u_i) \right\|_T \leq \|u_0\|_T + \frac{2\operatorname{card}(\Delta)}{2r} + \frac{\operatorname{card}(\Gamma)}{2r}$$

$$= 1 + \frac{\operatorname{card}(\Delta)}{r} + \frac{r - \operatorname{card}(\Delta)}{2r}$$

$$\leq \frac{3}{2} + \frac{\operatorname{card}(\Delta)}{2r} \leq \frac{7}{4}.$$

We finally get $\left\| u_0 + \frac{1}{r} \sum_{i=1}^{r} u_i \right\|_T \leq \frac{7}{4}$ as stated. ∎

We are now able to state the desired result.

Proposition 6.3. T *does not contain* ℓ_1.

Proof. Assume that T does contain ℓ_1. Then by a result due to James [91], for any $0 < \epsilon < 1$ there exists a sequence (u_i) of consecutive blocks on the unit ball of T with

$$(1 - \epsilon) \sum_i |a_i| \leq \left\| \sum_k a_i u_i \right\|_T \leq \sum |a_i| \qquad (6.6).$$

Let n_0 be the first element of the support of u_0 and set $r = 2n_0$. Then by (6.5) we have

$$\|u_0 + \frac{1}{r} \sum_{i=1}^{r} u_i\|_T \leq \frac{7}{4}.$$

On the other hand, (6.6) implies that

$$(1 - \epsilon)(1 + \frac{1}{r} \sum_{i=1}^{r} 1) = (1 - \epsilon) \cdot 2 \leq \frac{7}{4},$$

which yields a contradiction, since ϵ was arbitrary. ∎

As an application we get the following result.

Theorem 6.4. *The Tzirelson space* T *is reflexive. Moreover, the fundamental basis of any spreading model of* T *is equivalent to the canonical basis of* ℓ_1.

Proof. Since T contains neither c_0 or ℓ_1 and has an unconditional basis, we deduce, by Theorem 0.13 of Chapter 0, that T is reflexive. To complete the proof let us remark that any spreading model of T generated by any sequence can also be generated by a sequence of consecutive blocks. Thus the conclusion will follow from the inequalities (6.4). ∎

Since the spreading models of T are known, it is natural to ask what the spreading models of its dual T^* are. To answer this question, let us consider the biorthogonal system (e_i^*) associated with the canonical basis (e_i) of T. It is clear that (e_i^*) is a Schauder basis of T^*. Let $(a_i)_{1 \leq i \leq 2n}$ be a sequence of consecutive blocks with respect to (e_i^*) in the unit ball of T^*. By duality one can deduce from (6.4) that

$$\max_{n+1 \leq i \leq 2n} |a_i| \leq \Big\| \sum_{i=n+1}^{2n} a_i e_i^* \Big\|_{T^*} \leq 2 \max_{n+1 \leq i \leq 2n} |a_i|.$$

Therefore, as in the proof of Theorem 6.4, we can deduce that the fundamental basis of any spreading model of T^* is equivalent to the canonical basis of c_0.

In addition one can deduce from (6.4) that ℓ_1 is finitely representable in T. This implies that T is not super-reflexive, and therefore neither is T^*. Recall that a Banach space X is said to be *B-convex* if ℓ_1 is not finitely representable in X. (For more on B-convexity, see [19, 66, 150].)

By Theorem 4.8, we can conclude that T^* has the alternate Banach-Saks property.

Let us consider another example of a separable Banach space, one of whose interesting properties is that it cannot be embedded in a space with an unconditional basis. This example, which is due to James [92], played an important role in the development of Banach space theory and still remains a source of inspiration of many constructions.

Let $(x_n) \in \mathbf{R}^{(\mathbf{N})}$ and define $\|(x_n)\|_J$ by

$$\|(x_n)\|_J = \sup \frac{1}{\sqrt{2}} \Big\{ (x_{p_1} - x_{p_2})^2 + \cdots \\ + (x_{p_{n-1}} - x_{p_n})^2 + (x_{p_n} - x_{p_1})^2 \Big\}^{\frac{1}{2}}, \tag{6.8}$$

where the supremum is taken over all positive integers n and all increasing sequences of positive integers $(p_1, \ldots, p_n)$.

Definition 6.5. The *James space* J is the completion of $\mathbf{R}^{(\mathbf{N})}$ with respect to the norm $\| \cdot \|_J$.

We obviously have $x = (x_n) \in J$ if and only if $x \in c_0$ and $\|x\|_J < \infty$. Also, one can prove that the canonical basis of $\mathbf{R}^{(\mathbf{N})}$ is a bimonotone Schauder basis of J which is shrinking. One can use Theorem 0.8 of Chapter 0 to show that $(x_n) \in J^{**}$ if and only if $\|(x_n)\|_J < \infty$ and

$$\|(x_n)\|_{J^{**}} = \|(x_n)\|_J, \tag{6.9}$$

where by $\| \cdot \|_{J^{**}}$ we mean the bidual norm associated with $\| \cdot \|_J$. It is easy to see that when $\|(x_n)\|_J < \infty$, the sequence (x_n) is Cauchy and therefore converges. So the Banach space J^{**} is the linear span of J and the vector $(1, 1, 1, \ldots)$. Thus J is isometric to a maximal closed proper subspace of J^{**}. Now consider $T : J \to J^{**}$, defined by $T((x_n)) = (x_n - x_1)$. Using (6.9) one can check that T is an isometry which is also onto; this means that J and J^{**} are isometric.

Remark 6.6. The term $(x_{p_n} - x_{p_1})$ in formula (6.8) is introduced to enable T to be an isometry rather than just an isomorphism. Some authors define $\| \cdot \|_J$ without this term and the Banach space J_0 obtained is also called the James space. It is not hard to show that J and J_0 are isomorphic and $d(J, J_0) = \sqrt{2}$.

The James space J was used to disprove several long-standing conjectures in Banach space theory (for example, "a Banach space X is reflexive whenever X^{**} is separable" is _not_ true). Many other applications of J are given in [4, 5, 20, 41, 44, 93, 140].

We will wind up this section with the following technical lemma which will be of use in the next chapter.

Proposition 6.7.

(i) Let $F = \{i : a \leq i \leq b\}$, where $a \leq b$ are arbitrary integers. If P_F is the natural projection associated with the canonical basis (e_n) of J, then

$$\|I - P_F\| \leq \sqrt{2}. \tag{6.10}$$

(ii) Let $u = \sum_{i=n_1}^{n_2} \beta_i e_i$ and $v = \sum_{i=n_3}^{n_4} \alpha_i e_i$, with $n_2 + 2 \leq n_3$; then

$$\|u + v\|_J \leq \sqrt{2}\|u - v\|_J.$$

Proof. Since the same technique can be used to prove both *(i)* and *(ii)*, we only prove *(i)*.

Let $x = (x_n) \in J$ with $\|x\|_J \leq 1$; we have

$$(I - P_F)(x) = \sum_{i<a} x_i e_i + \sum_{i>b} x_i e_i = \sum_j y_j e_j.$$

Let $(p_j)_{j \leq n}$ be a strictly increasing finite sequence of integers. Then two cases can occur:

Case 1. $(p_j) \cap F = \phi$. Then

$$\sum_{i=1}^{n-1}(y_{p_i} - y_{p_{i+1}})^2 + (y_{p_n} - y_{p_1})^2 \leq \|x\|_J^2 \leq 1.$$

Case 2. $(p_j) \cap F \neq \phi$. Then

$$\sum_{i=1}^{n-1}(y_{p_i} - y_{p_{i+1}})^2 + (y_{p_n} - y_{p_1})^2 = \sum_{i \leq j}(x_{p_i} - x_{p_{i+1}})^2 + x_{p_j}^2 + x_{p_k}^2 +$$

$$\sum_{i=k}^{n-1}(x_{p_i} - x_{p_{i+1}})^2 + (x_{p_n} - x_{p_1})^2$$

with $p_j \leq a \leq b \leq p_k$. But

$$\sum_{i \leq j}(x_{p_i} - x_{p_{i+1}})^2 + \sum_{i=k}^{n-1}(x_{p_i} - x_{p_{i+1}})^2 + (x_{p_n} - x_{p_1})^2 \leq \|x\|_J^2 \leq 1,$$

and $x_{p_j}^2 - x_{p_k}^2 \leq \|x\|_J^2 \leq 1$, because the sequence (x_n) belongs to c_0. Finally we get $\|(I - P_F)(x)\|_J^2 \leq 2$. ∎

Chapter 3

I. AN INTRODUCTION TO FIXED POINT THEORY

Perhaps the most frequently cited fixed point theorem in analysis is the "Banach contraction mapping principle," which states that if (M, d) is a complete metric space and T is a contraction mapping from M to itself (there exists $0 < k < 1$ such that $d(Tx, Ty) \leq kd(x, y)$ for all x, y), then T has a unique fixed point in M. Moreover, for each $x \in M$ the Picard iterates $(T^n(x))$ converge to the fixed point. This theorem has its origins in Euler and Cauchy's work [47] on the existence and uniqueness of a solution to the differential equation

$$\begin{cases} dy/dx = f(x, y) \\ y(x_0) = y_0 \end{cases}$$

when f is a continuously differentiable function. In 1877, Lipschitz [144] simplified Cauchy's proof using what we now know as the "Lipschitz condition." (We should note that, interestingly, the method of Cauchy-Picard in fact was used before Cauchy.)

In 1890 Picard [157] applied the method of iterations to ordinary equations as well as to a class of partial differential equations. The formulation of the theorem given above is due to Banach [12]. An interesting generalization of the Banach contraction principle was given by Ekeland [60]. For more on this theorem the reader can consult [88, 180].

The Lipschitz condition $k < 1$ is crucial even for the existence part of the result, but within more restrictive settings an amplified fixed point theorem exists for the case $k = 1$. Mappings which satisfy the condition for $k = 1$ are known as *nonexpansive*, and the theory of nonexpansive mappings is fundamentally different from that of contraction mappings. For example, even if a nonexpansive mapping T has a nonempty set of fixed points Fix(T), the Picard iterates may fail to converge. Also, Fix(T) need not contain just one point.

Before we state the fixed point problem in Banach spaces, let us discuss the linear case, which is where the whole theory originated. Possibly the most important result in this case is the theorem of Brouwer [28, 29], which says that any continuous mapping taking the closed unit ball of $\mathbf{R}^n$ to itself has a fixed point. This result was previously known to Poincare [159] in an equivalent form. One interesting proof which is not based on algebraic topology is given by Scarf [168]. The underlying causes behind Brouwer's theorem are the compactness and convexity of the unit ball of $\mathbf{R}^n$. Thus in [170, 171] Schauder extended Brouwer's theorem to obtain the same conclusion for any compact convex set in any linear topological space which is locally convex.

Combining the two fundamental results described above, we come to the following basic problem.

Problem. Given a Banach space X and a nonempty, closed, bounded, convex subset K of X, what types of conditions on K (or X) will guarantee the existence of fixed points for every nonexpansive mapping of K onto itself?

Definition 1.1. A bounded, closed, convex subset K of a Banach space X is said to have the _fixed point property_ (f.p.p.) if every nonexpansive mapping T taking K to itself has a nonempty fixed point set (Fix(T) $\neq \phi$).

Recall that $T : K \rightarrow K$ is nonexpansive if

$$\|Tx - Ty\| \leq \|x - y\|$$

for all $x, y \in K$.

As the compactness assumption in Brouwer's theorem is so fundamental, one might naturally conjecture that any nonempty weakly compact convex subset of a Banach space has f.p.p. This question was closed in 1980, when Alspach [2] (also see [172, 178]) gave an example of a weakly compact convex subset of $L^1[0, 1]$ with an isometry which lacks a fixed point.

Let us add that the problem, as stated above, originated in four papers which appeared in 1965. In the first of these, Browder [30] proved that the problem has a positive answer when X is a Hilbert space. Almost immediately thereafter, both Browder [31] and Göhde [79] extended this conclusion to Banach spaces which are uniformly convex. At the same time, Kirk [114] observed that the presence of a geometric property called "normal structure" guarantees that weakly compact convex subsets have the fixed point property. See [75, 88, 112, 113, 163, 164, 180].

II. BASIC DEFINITIONS AND RESULTS

Let X be a Banach space and C a nonempty, weakly compact, convex subset of X. Assume that C does not have the fixed point property; therefore there exists a nonexpansive mapping $T : C \to C$ with an empty fixed point set.

Set

$$\mathcal{F} = \{K \subset C : K \text{ is nonempty, closed, convex, and}$$
$$\text{invariant under } T, \text{i.e. } TK \subset K\} \qquad (2.1)$$

Clearly $\mathcal{F}$ is a nonempty family, since $C \in \mathcal{F}$. It is easy to see that any decreasing chain of elements in $\mathcal{F}$ has a nonempty intersection, because C is weakly compact, which belongs to $\mathcal{F}$. Therefore we can use Zorn's lemma to demonstrate the existence of minimal elements.

Definition 2.1. A convex set K is said to be *minimal* for T if K is a minimal element of $\mathcal{F}$.

Clearly any set K which is minimal for T consists of more than one point, since otherwise T would have a fixed point, which contradicts our assumption. Kirk [114] was the first, in 1965, to investigate the structure of these minimal objects. No further progress was made until 1975, when both Goebel and Karlovitz published new information.

In this section we will give some basic results; these will be wrapped up in the following sections. For the rest of this section, K will be a minimal set for T.

II. Basic Definitions and Results

Proposition 2.2. *The following holds:*

$$\overline{\text{conv}}(TK) = K. \qquad (2.2)$$

Proof. Let $K_0 = \overline{\text{conv}}(TK)$; this is a closed, nonempty, convex subset of K since $TK \subset K$. Hence $TK_0 \subset TK \subset K_0$, so K_0 is invariant under T. Therefore $K_0 \in \mathcal{F}$ and since K is minimal we get $K_0 = K$. ∎

The next proposition gives an interesting result concerning minimal sets (see [149]).

Proposition 2.3. *Let $\alpha : K \to \mathbf{R}_+$ be a lower semi-continuous convex function. Assume that*

$$\alpha(Tx) \leq \alpha(x) \quad \textit{for all } x \in K. \qquad (2.3)$$

Then α is a constant function.

Proof. Let $x_0 \in K$ be fixed. Define

$$K_0 = \{x \in K : \alpha(x) \leq \alpha(x_0)\}.$$

K_0 is a closed, convex subset of K, since α is a lower semi-continuous convex function. Our assumption on α implies that K_0 is invariant under T, and since $x_0 \in K_0$, we deduce (by minimality of K) that $K_0 = K$. Therefore $\alpha(x) \leq \alpha(x_0)$ for all $x \in K$. But since x_0 was arbitrary, this completes the proof. ∎

In the following theorem we make use of Propositions 2.2 and 2.3 to obtain some properties of minimal sets.

Theorem 2.4. *The minimal set K is diametral, i.e.*

$$\sup_{y \in K} \|x - y\| = \text{diam}\, K \text{ for all } x \in K.$$

Proof. Set $\alpha(x) = \sup\{\|x - y\| : y \in K\}$. Then α is a continuous, convex function. If $x \in K$, then $K \subset B(x, \alpha(x))$, where $B(x, r)$ is the closed ball centered at x with radius r; since T is nonexpansive we deduce that $TK \subset B(Tx, \alpha(x))$. By proposition 2.2 $K = \overline{\text{conv}}TK \subset B(Tx, \alpha(x))$. This obviously implies that $\alpha(Tx) \leq \alpha(x)$. Hence α satisfies the conditions of Proposition 2.3, so that α is a constant. Say $\alpha(x) = \alpha$ for all $x \in K$. Since $\sup\{\|x - y\| : x, y \in K\} = \text{diam}K$, it follow that $\alpha = \text{diam}K$. ∎

Let X be a Banach space. For any bounded subset A of X, we define

$$r(x, A) = \sup\{\|x - y\| : y \in A\} \quad \text{for any } x \in A;$$
$$R(A) = \inf\{r(x, A) : x \in A\};$$
$$\delta(A) = \sup\{r(x, A) : x \in A\} = \text{diam}A;$$
$$C(A) = \{x \in A : R(A) = r(x, A)\}.$$

The positive number $R(A)$ is called the *Chebyshev radius* of A, and the set $C(A)$ is called the *Chebyshev center* of A. It should be clear that for any $x \in A$ we have

$$R(A) \leq r(x, A) \leq \delta(A). \tag{2.4}$$

In general there is no reason for $C(A)$ to be nonempty, except when A is a weakly compact nonempty convex set. A point $x \in A$ is called a *diametral point* if $r(x, A) = \delta(A)$, and the set A is called diametral if it consists only of diametral points. That is, A is diametral if and only if $C(A) = A$, which is also equivalent to $R(A) = \delta(A)$ by (2.4).

Normal structure, alluded to above, is a property which prohibits the existence of diametral convex sets.

Definition 2.5. A convex subset K of X is said to have *normal structure* if each bounded convex subset S of K with at least two points contains a non-diametral point. The Banach space X is said to have normal structure if each convex subset has normal structure.

The concept of normal structure was introduced in 1948 by Brodskii and Milman [27], who used it to prove that a weakly compact convex set which has normal structure contains a point which is fixed under every isometry from the set onto itself. (We will discuss further properties of normal structure at the end of this section.)

II. Basic Definitions and Results

As we said, a nonempty bounded set A has a non-diametral point if and only if $R(A) < \delta(A)$. In fact it may happen that there exists an $\alpha < 1$, independent of A, with

$$R(A) \leq \alpha \delta(A). \tag{2.5}$$

Definition 2.6. A Banach space X is said to have *uniform normal structure* if there exists $0 < \alpha < 1$ such that (2.5) holds for any nonempty bounded convex subset A of X.

Maluta [147] (see also [8]) has proved that uniform normal structure implies reflexivity. A similar result for metric spaces is given in [101]. It is still unknown whether uniform normal structure implies super-reflexivity (see [3, 147]). However, using Zizler's renorming technique [197], one can find an equivalent norm on c_0 (or more generally on any separable Banach space) such that c_0 has normal structure with respect to the new norm. Therefore normal structure is not a super-property.

For a while it was thought that any reflexive Banach space has normal structure. James [22] disproved this statement by renorming ℓ_2 with

$$\|x\|_\beta = \max\{\|x\|_{\ell_2}, \beta\|x\|_\infty\} \tag{2.6}$$

where $x \in \ell_2$ and $\beta > 0$. Write $X_\beta = (\ell_2, \|\cdot\|_\beta)$; then James proved that $X_{\sqrt{2}}$ fails to have normal structure. In fact, one can prove that X_β has normal structure if and only if $\beta < \sqrt{2}$. In [11], where Baillon and Schönberg have studied the spaces X_β intensively, the concept of *asymptotic normal structure* is introduced. A Banach space has asymptotic normal structure if, for each nonempty bounded closed convex subset K, and for each sequence (x_n) in K with $\|x_n - x_{n+1}\| \to 0$, there is a point $x \in K$ such that

$$\liminf_n \|x_n - x\| < \delta(K).$$

Baillon and Schönberg proved that X_β has asymptotic normal structure if and only if $\beta < 2$. Therefore asymptotic normal structure does not necessarily imply normal structure, although the converse is true.

On normal structure and related concepts in Banach spaces, one can consult [75, 77, 78, 86, 100, 104, 106, 133, 152, 188, 194]. In the next result we discuss the connection between normal structure and the fixed point property.

Theorem 2.7. *Let X be a Banach space. If C is a nonempty weakly compact convex subset of X, and C has normal structure, then C has f.p.p.*

Proof. Assume to the contrary that C is such a set which fails to have f.p.p. Therefore C contains a diametral convex subset by Theorem 2.4, contradicting the assumption. ∎

This theorem was first proven by Kirk [114]. Baillon and Schönberg [11] strengthened this result by proving that asymptotic normal structure implies f.p.p. It was proven directly in [11] that X_2 has f.p.p.; in the case $\beta > 2$, the situation was unclear. We will see that X_β has f.p.p. for any $\beta > 0$.

It was unknown whether normal structure and the fixed point property are equivalent until 1975, when Karlovitz [96] proved that $X_{\sqrt{2}}$ has f.p.p. He obtained this result using a new property satisfied by minimal convex sets which was discovered independently by Göebel [73]. Before we go into this result, let us recall some basic facts about nonexpansive mappings.

Let K be a nonempty, bounded, closed, convex subset of a Banach space X. Let $T : K \to K$ be a non-expansive mapping. Fix $0 < \epsilon < 1$ and $z \in K$, and consider the mapping $T_\epsilon : K \to K$ defined by

$$T_\epsilon(x) = \epsilon z + (1 - \epsilon)T(x) \tag{2.7}$$

for all $x \in K$. T_ϵ is clearly a contraction mapping, and therefore has a unique fixed point x_ϵ in K. Now one can easily prove that $\|x_\epsilon - T(x_\epsilon)\| \leq \epsilon\delta(K)$, so that $\inf\{\|x - T(x)\| : x \in K\} = 0$. (When T is not nonexpansive, we generally have $\inf\{\|x - T(x)\| : x \in K\} > 0$. The evaluation of this infimum is connected to the so-called minimal displacement. See [68, 74, 76].)

Definition 2.8. A sequence (x_n) satisfying $\|x_n - T(x_n)\| \to 0$ as $n \to \infty$ is called an *approximate fixed point sequence*, in short an a.f.p.s.

We proved that we can always find an approximate fixed point sequence in K provided that T is nonexpansive. Now the property

II. Basic Definitions and Results

satisfied by minimal sets, as proven by Karlovitz and Göebel, can be stated.

Proposition 2.9. *Let K be a minimal set for T. Then for any a.f.p.s. (x_n) in K, we have*

$$\lim_{n \to \infty} \|x_n - x\| = \delta(K) \tag{2.8}$$

for all $x \in K$.

Proof. Set $\alpha(x) = \lim_{\mathcal{U}} \|x_n - x\|$, where $\mathcal{U}$ is an ultrafilter on $\mathbf{N}$. The function α is well defined because (x_n) is bounded, and α is clearly continuous and convex. Since (x_n) is an a.f.p.s. for T, it follows that $\alpha(T(x)) \leq \alpha(x)$ for any $x \in K$. Therefore α satisfies the conditions of Proposition 2.3, so α must be a constant function, say $\alpha(x) = \alpha$. Using the weak compactness of K, we can deduce that the weak limit of (x_n) over $\mathcal{U}$ exists in K. Put $z = w - \lim_{\mathcal{U}}(x_n)$, where $w - \lim$ is the weak limit. Since the norm is weak lower semi-continuous, we obtain

$$\|z - x\| \leq \lim_{\mathcal{U}} \|x_n - x\| = \alpha$$

for any $x \in K$. Proposition 2.4 implies that $\alpha = \operatorname{diam} K$. Since $(\|x_n - x\|)$ has a unique cluster point, it is convergent. This completes the proof. ∎

We saw that the fixed point problem in Banach spaces can be seen as a combination of the Brouwer theorem and the Banach contraction theorem. Since the compactness assumption in Brouwer's theorem is taken with respect to the strong topology, it was natural to consider, as a variant and a generalization of the problem, compactness with respect to other topologies, e.g. the weak*-topology in dual Banach spaces. In [110] the authors consider another topology in Banach lattices which can be reduced to well-known topologies in certain cases. A surprising fact is that the fixed point problem behaves differently from one topology to another; for example, Lim [132] found an easy example of a weak*-compact closed convex set which fails to have the fixed point property. (Alspach's counterexample in [2] is very complicated.)

Let us say that a Banach space X has the $\mathcal{T}$-fixed point property if any $\mathcal{T}$-compact convex subset of X has f.p.p. Since we will consider the weak*-topology in dual Banach spaces later, let us note an important fact about duals. Let X be a dual Banach space and let X_* be a predual of X, i.e. the dual space of X_* is isometric to X. Then the couple (X_*, X) defines a weak*-topology on X usually denoted by $\sigma(X, X_*)$. The problems start when the predual space X_* is not unique. Then there is no reason why the weak*-topology should be unique. For example, it is not hard to show that the dual space of c, the space of convergent sequences, is ℓ_1. On ℓ_1, the topologies $\sigma(\ell_1, c)$ and $\sigma(\ell_1, c_0)$ are different, since in [102] it is shown that ℓ_1 fails to have the $\sigma(\ell_1, c)$-fixed point property but does have the $\sigma(\ell_1, c_0)$-fixed point property.

For the time being, the weak*-topology has to be understood with respect to a fixed predual. A specific remark will be made whenever there is a possible ambiguity.

Let C be a nonempty, weak*-compact, convex subset of a dual space X, and let $T : C \to C$ be a nonexpansive mapping with empty fixed point set $\mathrm{Fix}(T)$. Then by once again applying Zorn's lemma, it is possible to find a minimal nonempty weak*-closed convex subset K of C which is invariant under T. One can then easily prove the following.

Proposition 2.10 Let $\alpha : K \to \mathbf{R}_+$ be a weak*-semicontinuous convex function which satisfies $\alpha(Tx) \leq \alpha(x)$ for all $x \in K$. Then α is constant.

Using the easy fact that $\overline{\mathrm{conv}}^*(TK) = K$, one can deduce from Proposition 2.10 that K is diametral. As for the weak*-topology, one can define the weak*-normal structure and obtain an equivalent statement to Theorem 2.7. More on weak*-normal structures is given in [104, 188].

A natural question is whether the conclusion of Proposition 2.9 holds for the K we are considering. The answer is not easy, since it is not clear whether the functions $\alpha(x) = \lim_{\mathcal{U}} \|x_n - x\|$, considered in the proof of Proposition 2.9, are weak*-lower semi-continuous. These functions are called *types* on X. More on the properties of types can be found in [18, 26, 82, 124]. Little is known about the general question, but in certain special cases we have a positive answer. Indeed, let X ba a Banach space with a Schauder basis (e_n). We sill say that (e_n) is strongly bimonotone if $\|P_F\| = \|I - P_F\|$ for every segment

F in $\mathbb{N}$, where P_F denotes the natural projection on (e_n) associated with F. In [102] the following is proven.

Proposition 2.11. *Let X be a Banach space with a shrinking Schauder basis (e_n). Assume that (e_n) is strongly bimonotone and let K be a minimal weak*-compact convex subset of X^* for a nonexpansive mapping T. Then for any a.f.p.s. (x_n) of T (in K) and for all $x \in K$,*
$\lim_{n \to \infty} \|x_n - x\| = \operatorname{diam} K$ *holds.*

Proof. Let K be a minimal weak*-compact convex subset of X^*, for a nonexpansive mapping T. Let (x_n) be any a.f.p.s. of T and let $x \in K$; and let s be a cluster point of $(\|x_n - x\|)$. Then there exists a subsequence (x'_n) of (x_n) so that

$$\lim_{n \to \infty} \|x'_n - x\| = s. \tag{2.9}$$

Let $\mathcal{U}$ be an ultrafilter on $\mathbb{N}$ and x_0 the weak*-limit of (x'_n) over $\mathcal{U}$. Since (e^*_i) is a basis of X^*, by applying Proposition 0.14 of Chapter 0 one can find a sequence (u_n) of successive blocks (related to (e^*_i)) and a subsequence (x''_n) of (x'_n) so that

$$\lim_{n \to \infty} \|x''_n - u_n - x_0\| = 0. \tag{2.10}$$

Set $r(y) = \lim_{n,\mathcal{U}} \|x''_n - y\|$ for any $y \in K$. Let $y_0 \in K$ and consider the set

$$K_0 = \{y \in K : r(y) \le r(y_0) \le r_0\}.$$

K_0 is a nonempty convex subset of K. Since (x''_n) is an a.f.p.s. for T, we deduce that $r(Ty) \le r(y)$ for all $y \in K$. Therefore K_0 is invariant under T.

Now let us prove that K_0 is weak*-closed. Let (y_n) be a sequence weak*-convergent to y, with $y_n \in K_0$ for all n. Again by Proposition 0.14, there exists a sequence (v_n) of successive blocks and a subsequence (y'_n) of (y_n) such that

$$\lim_{n \to \infty} \|y'_n - v_n - y\| = 0. \tag{2.11}$$

Since r is continuous, we may assume that y and x_0 have finite support. And since the basis (e^*_i) is strongly bimonotone, for m large enough, we have

$$\|x_0 + u_m - y\| \le \|x_0 + u_m - y - v_n\|.$$

Hence

$$\lim_{m,\mathcal{U}} \|x_0 + u_m - y\| \leq \lim_{m,\mathcal{U}} \|x_0 + u_m - y - v_n\|. \qquad (2.12)$$

By (2.12) we have

$$r(y) \leq r(y + v_n),$$

and since r is continuous, we have from (2.11) that

$$r(y) \leq \lim_{n \to \infty} \inf\, r(y'_n) \leq r(y_0).$$

Therefore $y \in K_0$. It then follows from the minimality of K that $K = K_0$. In other words r is constant on K, i.e. $r(y) = r$ for all $y \in K$. Since the norm is weak*-lower semi-continuous, we deduce

$$\|y - x_0\| \leq \lim_{n \to \infty} \inf\, \|x''_n - y\| \leq r.$$

From Proposition 2.10, we get

$$\sup\{\|x_0 - y\| : y \in K\} = \mathrm{diam}\, K.$$

Hence $\mathrm{diam}\, K \leq r$, which implies $\mathrm{diam}\, K = r$. But since $r(x) = s$ that means $s = \mathrm{diam}\, K$, and the proof is complete. ∎

NOTES ON NORMAL STRUCTURE

For the sake of completeness, we give this note on normal structure properties. We start by recalling some basic definitions.

Definition (1). Let X be a Banach space and K a bounded convex subset of X. We say that K has *normal structure* (resp. *uniform normal structure*) if every nontrivial convex subset C of K contains a point x such that

$$r(x, C) = \sup\{\|x - u\| : y \in C\} < \mathrm{diam}\, C$$

(resp. $r(x, C) \leq \alpha\, \mathrm{diam}\, C$, where $\alpha \in (0, 1)$ and is independent of C).

We say X has normal structure (resp. uniform normal structure) if every bounded convex subset K has normal structure (resp. uniform normal structure with constant α independent of K).

Since we sometimes deal with weakly compact convex subsets, or weak*-compact convex subsets, of dual spaces, we introduce the following concept.

Definition (2). Let X be a (resp. dual) Banach space. We say that X has *weak normal structure* (resp. *weak*-normal structure*) if any weakly (resp weak*-) compact convex subset of X has normal structure.

For the interested reader, we note a list of works on normal structure, [77, 119, 152, 188]. Let us remark that Banach spaces which have uniform normal structure are reflexive [8, 147] (see [101] for an extension to metric spaces).

In the rest of this section we will use the notation n.s., w.n.s, and w*.n.s. To determine whether a given convex bounded subset K of X has n.s. is not an easy task. The first characterization of n.s. was given by Brodskii-Milman [27]; it is still widely used.

Theorem (3). *Let X be a Banach space and K a convex bounded subset of X. The following are equivalent:*

(i) K fails to have n.s.;

(ii) K contains a non-constant diametral bounded sequence, i.e.

$$\lim_{n \to \infty} d(x_{n+1}, \operatorname*{conv}_{i \le n}(x_i)) = \operatorname{diam}(x_i);$$

(iii) for every $\epsilon > 0$, K contains a non-constant bounded sequence (x_n) such that

$$\|x_{n+1} - x\| \ge \operatorname{diam}(x_i)(1 - \frac{1}{n^\epsilon})$$

for every $n \in \mathbf{N}$ and every $x \in \operatorname*{conv}_{i \le n}(x_i)$.

Landes [129, 130] has extended this characterization, enabling him to prove some nice structure theorems on spaces with normal structure.

1. Spaces with normal structure properties.

Our first example concerns finite dimensional spaces [22, 188].

Proposition (1). *Let X be a Banach space and K a compact convex subset of X. Then K has n.s., and therefore finite dimensional Banach spaces have n.s.*

This result is important since it shows that the normal structure property is weakly linked to any geometrical property. In the next example, we discuss a result of Zizler [197] and certain connected problems.

Definition (2). Let X ba a Banach space. Define the *modulus of uniform convexity* δ in the direction $z \in X$ ($\|z\| = 1$) by

$$\delta(z, \epsilon) = \inf\{1 - \|\frac{x+y}{2}\| : \|x\| \le 1, \|y\| \le 1, \text{ and}$$
$$x - y = \alpha z, \text{ with } \|x - y\| \ge \epsilon\}$$

for any $\epsilon \in [0, 2)$. X is said to be *U.C.E.D.* if $\delta(z, \epsilon) > 0$ always.

Zizler [197] (see also [54, 69]) has proved the following.

Proposition (3). *Let X be a Banach space which is U.C.E.D. Then any bounded convex subset of X has n.s.*

Zizler also proved that any separable Banach space X can be equipped with an equivalent norm which is U.C.E.D. Let us remark that for a while it was unknown whether any reflexive Banach space has an equivalent norm which is U.C.E.D. This problem was solved negatively by Kutzarova and Troyanski [125]. It is still unknown

Notes on Normal Structure

whether any reflexive Banach space has an equivalent norm which has n.s.

Let us add that Landes [130] characterized Banach spaces with symmetric Schauder bases (not necessarily countable) which can be renormed to have n.s. and those which can be renormed to be U.C.E.D. (For the definition of a symmetric Schauder basis one can consult [143].) In particular he proved that $c_0(I)$ can be renormed to have n.s. if and only if I is countable. Van Dulst [56] has also shown that every Banach space may be equivalently renormed so as to lack n.s.

Proposition (3) was generalized by some authors (see Smith [181], Khamsi [106]). It is of interest to note that Fakhouri [65] (see also Smith [181]) introduced a concept of uniform convexity in the direction of some subsets.

We now turn our attention to Banach spaces which are close to being uniformly convex. Recall that the definition of the modulus of uniform convexity of X is

$$\delta_X(\epsilon) = \inf\{1 - \|\frac{x+y}{2}\| : \|x\| \leq 1, \|y\| \leq 1, \text{ and } \|x - y\| \geq \epsilon\}$$

for any $\epsilon \in (0,2)$.

Definition (4). The *characteristic of uniform convexity* ϵ_0 of X is defined by

$$\epsilon_0(X) = \sup\{\epsilon : \delta_X(\epsilon) = 0\}.$$

X is said to be *U.C.* if $\epsilon_0(X) = 0$ and *uniformly non-square* if $\epsilon_0(X) < 2$.

James [90] proved that X is super-reflexive whenever $\epsilon_0(X) < 2$. In the case when $\epsilon_0(X) < 1$ we have a nice conclusion:

Proposition (5). *Suppose that* $\epsilon_0(X) < 1$. *Then for any bounded closed convex subset C of X, there exists $x \in C$ such that*

$$\sup\{\|x - y\| : y \in C\} \leq (1 - \delta_X(1))\text{diam}C,$$

which implies that X has uniform normal structure.

In order to give Baillon's result [9] on uniform smooth Banach spaces, we need the following definition.

Definition (6). Let X ba a Banach space. The *modulus of smoothness* ρ_X of X is defined by

$$\rho_X(\zeta) = \sup\{\frac{1}{2}[\|x + \zeta y\| + \|x - \zeta y\| - 2] : \|x\| \leq 1, \|y\| \leq 1\}$$

for every $\zeta > 0$. X is said to be *U.S.* if $\lim\limits_{\zeta \to 0} \frac{\rho_X(\zeta)}{\zeta} = 0$.

In [100], the following technical lemma is proven.

Lemma (7). *For any Banach space X,*

$$\lim_{\zeta \to 0} \frac{\rho_X(\zeta)}{\zeta} = \frac{1}{2}\epsilon_0(X^*).$$

From the proof given by Baillon in [9] and using this lemma, one can deduce the following result, which was stated by Turrett [194] and Khamsi [100].

Theorem (8). *Let X be a Banach space such that $\lim\limits_{\zeta \to 0} \frac{\rho_X(\zeta)}{\zeta} < \frac{1}{2}$; then X and X^* are super-reflexive and have super-n.s.*

Sullivan [187] (also see [23]) has generalized the concept of uniform convexity by introducing the *k*-U.C., for $k \in \mathbb{N}$.

Definition (9). The *modulus of k-uniform convexity δ_X^k* of a Banach space X is defined by

$$\delta_X^k(\epsilon) = \inf\{1 - \frac{\|x_1 + \ldots + x_{k+1}\|}{k+1} : \|x_i\| \leq 1 \text{ and}$$
$$V(x_1, \ldots, x_{k+1}) > \epsilon\}$$

Notes on Normal Structure

for all $\epsilon \in (0,2)$, where $V(x_i)$ is

$$\sup\left\{ \begin{vmatrix} 1 & 1 & \cdots & 1 \\ f_1(x_1) & f_1(x_2) & \cdots & f_1(x_{k+1}) \\ \vdots & \vdots & \ddots & \vdots \\ f_k(x_1) & f_k(x_2) & \cdots & f_k(x_{k+1}) \end{vmatrix} : f_i \in X^*, \|f_i\| \leq 1 \right\}.$$

X is said to be k-$U.C.$ whenever $\delta_X^k(\epsilon) > 0$ for every $\epsilon \in (0,2)$.

Amir [3] gave a statement equivalent to Proposition (5). First, define the characteristic of k-uniform convexity ϵ_0^k of X by

$$\epsilon_0^k = \sup\{\epsilon : \delta_X^k(\epsilon) = 0\}.$$

Proposition (10). *Let X be a Banach space and assume that $\epsilon_0^k(X) < 1$ for some $k \in \mathbb{N}$. Then X has uniform normal structure.*

On k-$U.C.$ and some structure results, one can consult [117, 118, 182, 189, 190, 191].

Recently, Smith and Turett [182], using the conclusion of Proposition (10), proved that uniform normal structure is not a self-dual property. This extends Bynum's result [38].

Another geometrical property connected to n.s. was introduced by Huff [86].

Definition (11). Let X be a Banach space. X is said to be *nearly uniformly convex* if for every $\epsilon > 0$ there exists $\delta(\epsilon) > 0$ such that if $\|x_n\| \leq 1$ and $\mathrm{spe}(x_n) = \inf\{\|x_n - x_m\| : n \neq m\} \geq \epsilon$, then $\mathrm{conv}(x_n) \cap B(0, 1 - \delta(\epsilon)) \neq \phi$.

Van Dulst and Sims [57] proved the following.

Proposition (12). *Every nearly uniformly convex Banach space has n.s.*

Let us remark that in all the propositions cited above, we have assumed that the entire space has some geometrical property. In [71,

104, 106, 189] it is shown that sometimes it is enough to assume that a finite codimensional subspace has the property.

We now turn to Opial's condition [154]. Notice that most of the above geometric properties are satisfied in a space which is uniformly convex. Here we have a condition which is not satisfied in L^p, for $p \geq 1$.

Definition (13). A Banach space X is said to satisfy *Opial's condition* if for every sequence (x_n) weakly convergent to w,

$$\liminf_{n \to \infty} \|x_n - w\| < \liminf_{n \to \infty} \|x_n - x\|$$

holds, for all $x \neq w$.

This definition is motivated by the fact that this property implies that the asymptotic center of a sequence coincides with its weak limit, which of course fails in L^p for $p \geq 1$ (and more generally in Orlicz spaces L^φ, see [127, 128]).

Opial's condition is connected to n.s. This fact was observed by Gossez and Lami-Dozo [81].

Proposition (14). *Every Banach space which satisfies Opial's condition has weak-normal structure.*

We add that spaces which satisfy Opial's condition not only have the fixed point property (in accordance with Kirk's theorem), but also satisfy the so-called demi-closedness principle.

Theorem (15). *Let X be a Banach space which satisfies Opial's condition and let K be a weakly compact convex subset of X, with a nonexpansive self-mapping $T : K \to K$. Then the mapping $I - T$ is demi-closed on K, i.e. if (u_n) is weakly convergent to u and $(u_n - Tu_n)$ converges strongly to w, then $u - Tu = w$.*

This is surprising since Browder [31] has noted that the demi-closedness principle holds in any uniformly convex space.

We now introduce another concept which makes sense in infinite dimensional spaces and which lead to generalizations of Proposition (1). Unfortunately, this concept did not have a great success,

65

as the other geometric properties did, despite the nice results obtained concerning the fixed point property and normal structure (see [14,78,173]).

Let X be a Banach space and A a bounded subset of X. *Kuratowski's measure of noncompactness* α of A is defined by

$$\alpha(A) = \inf\{d > 0 : A \text{ can be covered with a finite number of}$$
$$\text{sets of diameter smaller than } d\}.$$

Concerning the basic properties of $\alpha(\cdot)$, one can consult [1, 7, 14, 15, 78, 173].

Definition (16). Let X be a Banach space. The *modulus of noncompact convexity* of X is given by

$$\Delta_X(\epsilon) = \inf\{1 - \inf_{x \in A} \|x\| : A \text{ is a convex subset}$$
$$\text{of the unit ball with } \alpha(A) \geq \epsilon\}$$

for every $\epsilon \in (0, 2)$. The *characteristic of noncompact convexity* ϵ_1 of X is given by

$$\epsilon_1(X) = \sup\{\epsilon : \Delta_X(\epsilon) = 0\}.$$

This concept was introduced by Goebel and Sekowski [78], who proved the following (also see [173]).

Proposition (17). *If X is a Banach space with $\epsilon_1(X) < 1$, then X is reflexive and has n.s.*

2. Normal structure in spaces with bases.

One of the first criteria which implies n.s. in spaces with a Schauder basis was introduced by Gossez and Lami-Dozo [80].

Definition (1). Let X be a Banach space with a Schauder basis (e_n). We will say that (e_n) satisfies the property *G.L.D.* if there exists a strictly increasing sequence (n_k) of integers such that for

every $c > 0$, we can find $r = r(c) > 0$ such that for every k and every $x \in X$,
$$\|P_{n_k}(x)\| = 1 \quad \text{and} \quad \|(I - P_{n_k})(x)\| \geq c$$
implies that $\|x\| \geq 1 + r$ (where $x \in X$).

Gossez and Lami-Dozo [80] proved the following.

Proposition (2). *Let X be a Banach space with a Schauder basis satisfying G.L.D. Then any weakly compact convex subset of X has n.s.*

The G.L.D. property was very useful in studying n.s. in certain spaces, such as Orlicz's sequence spaces [58] (see also [184]). A generalization of G.L.D. was given by Bynum [37].

In [104] the author associated to any Banach space with a Schauder basis (and more generally with a F.D.D.) an easily calculable constant which in related to n.s.

Definition (3). Let X be a Banach space with a Schauder basis. Define $\beta_p(X)$, for $p \in [1, \infty)$, to be the infimum of the set of numbers λ such that
$$(\|x\|^p + \|y\|^p)^{\frac{1}{p}} \leq \lambda \|x + y\|$$
for every $x, y \in X$ which satisfy $\text{supp}(x) < \text{supp}(y)$. (We mean any $i \in \text{supp}(x)$ is less than any $j \in \text{supp}(y)$.)

Proposition (4). *Let X be a Banach space with a finite codimensional subspace Y such that $\beta_p(Y) < 2^{\frac{1}{p}}$ for some $p \in [1, \infty)$. Then X has weak-n.s.*

In [104], an elementary proof to Bynum's [39] result on $\ell_{p,1}$ is given. Bynum's original proof uses three coefficients associated to each Banach space, which are closely related to n.s.

Definition (5). (1) The *normal structure coefficient* of X, denoted by $N(X)$, is defined by
$$N(X) = \inf\{\frac{\delta(K)}{R(K)} : K \text{ is a closed convex subset of}$$
$$X \text{ with more than one point}\},$$

Notes on Normal Structure

where $R(K) = \inf_{x \in K}\{\sup\{\|x - y\| : y \in K\}\}$.

(2) The *bounded sequence coefficient* of X, denoted by $BS(X)$, is the supremum of all numbers M such that for each bounded sequence (x_n) there is a $y \in \overline{\text{conv}}(x_n)$ such that

$$M \cdot \limsup_{n \to \infty} \|x_n - y\| \leq A(x_n),$$

where $A(x_n) = \lim_{n \to \infty}(\sup\{\|x_m - x_k\| : m, k \geq n\})$.

(3) The *weakly convergent sequence coefficient* of X, denoted $WCS(X)$, is defined like $BSC(X)$, replacing "bounded" by "weakly convergent."

Bynum [39] proved:

Proposition (6). *Let X be a Banach space. Then $1 \leq N(X) \leq BSC(X) \leq WCS(X) \leq 2$, and if one of these coefficients is greater than 1 then X has n.s.*

3. Some generalizations.

One of the first generalizations of n.s. was introduced by Belluce and Kirk [21] in order to obtain a fixed point theorem for any commutative family of nonexpansive mappings. First let us define some notation; for any subset K and any bounded subset H of a Banach space X, set

$$r(x, H) = \sup\{\|x - y\| : y \in H\};$$
$$r(H, K) = \inf\{r(x, H) : x \in K\};$$
$$C(H, K) = \{x \in K : r(x, H) = r(H, K)\}.$$

The set C is called the *Chebyshev center* of H with respect to K.

Definition (1). Let K be a bounded closed convex subset of a Banach space X. We say that K has *complete normal structure* (in short, c.n.s.) if every convex subset W of K which contains more than one point satisfies the following condition (BK):

For every decreasing net $\{W_\alpha : \alpha \in A\}$ of subsets of W which satisfy $r(W_\alpha, W) = r(W, W), \alpha \in A$, it is the case that the closure of $\bigcup\limits_{\alpha \in A} \mathcal{C}(W_\alpha, W)$ is a nonempty proper subset of W.

It is clear that c.n.s. implies n.s., but it was unknown whether the converse was true until Lim [133] gave a positive solution.

Another generalization was given by Baillon and Schoneberg [11].

Definition (2). The Banach space X is said to have *asymptotic normal structure* (a.n.s.) if for every bounded closed convex susbet K of X with positive diameter, and for any sequence (x_n) in K satisfying $\|x_{n+1} - x_n\| \to 0$ as $n \to \infty$, there exists $x \in K$ wuch that

$$\liminf_{n \to \infty} \|x_n - x\| < \operatorname{diam} K.$$

For more on a.n.s. one can consult [39, 148, 193].

4. Normal structure in metric spaces.

It is tempting to try to generalize the normal structure property to metric spaces, in order to obtain a similar result to Kirk's theorem. One of the first generalizations was given by Kijima and Takahashi [111] (also see [146, 192]). Their work didn't have a big success in application since their definition of convexity in metric spaces was too constraining. It seems that Penot [156] was the first who freed the n.s. property from linear convexity. We will now give Penot's formulation and some of its applications.

Definition (1). Let (M, d) be a metric space. A nonempty family $\mathcal{F}$ of subsets of M is called a *convexity structure* if it is stable under arbitrary intersections.

In the following we will always assume that any convexity structure contains the closed balls.

69

Notes on Normal Structure

Definition (2). Let $\mathcal{F}$ be a convexity structure on (M, d).

(i) We will say that $\mathcal{F}$ is *compact* if any family $(C_\alpha)_{\alpha \in \Gamma}$ of elements of $\mathcal{F}$ with the finite intersection property (every intersection of finitely many C_α's is nonempty) has a nonempty intersection;

(ii) $\mathcal{F}$ is called *normal* (resp. *uniformly normal*) if for every bounded $A \in \mathcal{F}$ with positive diameter, there exists $x \in A$ such that $\sup\{d(x, y) : y \in A\} < \operatorname{diam}A$ (resp. $\sup\{d(x, y) : y \in A\} \leq c\operatorname{diam} A$, for some $c \in (0, 1)$ which is independent of A).

We remark that the "compactness" of Definition (2) is equivalent to weak-compactness in the linear case.

The most interesting application of these concepts was given by Sine [175] and Soardi [185]. Indeed, let M be the unit ball of ℓ_∞ and consider the convexity structure $\mathcal{A}(M)$ defined by

$$\mathcal{A}(M) = \{A \subset M : A \text{ is an intersection of closed balls}\}. \quad (1)$$

Since closed balls are weak*-compact, one can deduce that $\mathcal{A}(M)$ is compact. Sine and Soardi proved that for any $A \in \mathcal{A}(M)$ consisting of more than one point, there exists $x \in A$ such that

$$\sup\{d(x, y) : y \in A\} = \frac{1}{2}\operatorname{diam}(A).$$

This implies that $\mathcal{A}(M)$ is uniformly normal. Notice that if one takes $\mathcal{F}$ to be the family of convex subsets of M, $\mathcal{F}$ is neither compact nor normal.

This example is a particular case of a category of metric spaces introduced by Aronszajn and Panitchpakdi [6].

Definition (3). Let (M, d) be a metric space. M is called *hyperconvex* if for every family $(x_\alpha)_{\alpha \in \Gamma}$ in M and every $(r_\alpha)_{\alpha \in \Gamma}$ in $\mathbf{R}$ such that for any α and β such that

$$d(x_\alpha, x_\beta) \leq r_\alpha + r_\beta,$$

we have $\bigcap_{\alpha \in \Gamma} B(x_\alpha, r_\alpha) \neq \phi$.

The family $\mathcal{A}(M)$ defined by (1) is compact and uniformly normal for *any* M which is hyperconvex. More on hyperconvex spaces is given in [10, 87, 109, 134, 135, 136, 161, 162, 176, 177].

Since these concepts were introduced in order to get a fixed point result, let us give the following theorem [156].

Theorem (4). *Let (M, d) be a bounded metric space. Assume that (M, d) has a convexity structure $\mathcal{F}$ which is compact and normal. Then any nonexpansive mapping $T : M \to M$ has a fixed point.*

We remark that the compactness assumption can be weakened (see [72, 156]).

One also finds in Penot's formulation an advantage to generalizing these concepts to more abstract structures (going beyond metric spaces). An interesting illustration of this if given in the excellent paper by Jawhari, Misane, and Pouzet [95]. For more on this problem one can consult [49, 107, 108, 160].

III. BASIC RESULTS IN ULTRAPRODUCT LANGUAGE

Let C be, as usual, a non-empty, convex, weakly-compact subset of a Banach space X. Suppose that C does not have the fixed point property; therefore there exists a nonexpansive mapping $T : C \to C$ whose fixed point set Fix(T) is empty. As before, let K be a minimal set for T. From now on $\mathcal{U}$ will stand for a non-trivial ultrafilter on the set of positive integers. In the ultrapower $(X)_\mathcal{U}$ of X, define $\tilde{K}$ and $\tilde{T}$ as in section V of Chapter 2, i.e.

$$\tilde{K} = \{\tilde{x} \in (X)_\mathcal{U} : \text{there exists a representative } (x_n) \text{ of } \tilde{x} \text{ with}$$
$$x_n \in K \text{ for any } n \geq 0\},$$

and

$$\tilde{T}\tilde{x} = (\widetilde{Tx_n}) \text{ for any } \tilde{x} \in \tilde{K}.$$

$\tilde{K}$ is a bounded, closed, convex subset of $(X)_\mathcal{U}$ which is invariant under $\tilde{T}$. Let us remark that $\tilde{K}$ is *not* a minimal set for $\tilde{T}$. Indeed, we proved that T has an a.f.p.s. (x_n) in K, i.e. $\|x_n - Tx_n\| \to 0$ as $n \to \infty$. Let $\tilde{x} = (\widetilde{x_n})$; then $\tilde{x} \in \tilde{K}$ and we clearly have $\tilde{T}\tilde{x} = \tilde{x}$. This means that the set Fix$(\tilde{T})$ is not empty, hence $\tilde{K}$ cannot be a minimal set for $\tilde{T}$. Recall that if $\tilde{x} = (\widetilde{x_n}) \in$ Fix$(\tilde{T})$, then there exists a subsequence (x'_n) of (x_n) which is an a.f.p.s. for T. On the other hand, if $x \in K$ and $\tilde{x} \in \tilde{K}$, and we assume that $\tilde{x} \in \text{Fix}(\tilde{T})$, then the conclusion of Proposition 2.9 implies that

$$\|\tilde{x} - x\|_{(X)_{\mathcal{U}}} = \text{diam} K. \qquad (3.1)$$

The properties of $\tilde{K}$ are summarized in the following theorem.

Theorem 3.1. *The following hold.*
  (i) $\text{diam} \tilde{K} = \text{diam} K = \text{diam Fix}(\tilde{T})$.

  (ii) K, $\tilde{K}$ and $\text{Fix}(\tilde{T})$ are diametral.

  (iii) $\text{Fix}(\tilde{T})$ is metrically convex, i.e. for any $\tilde{x}$, $\tilde{y}$ in $\text{Fix}(\tilde{T})$ and any $0 \leq \alpha \leq 1$, there exists $\tilde{z} \in \text{Fix}(\tilde{T})$ so that

$$\|\tilde{x} - \tilde{z}\| = (1 - \alpha)\|\tilde{x} - \tilde{y}\|$$

and

$$\|\tilde{y} - \tilde{z}\| = \alpha\|\tilde{x} - \tilde{y}\|$$

hold.

  (iv) Let $(\tilde{w}_n)$ be an a.f.p.s. for $\tilde{T}$ in $\tilde{K}$; then for any $x \in K$ we have

$$\lim_{n \to \infty} \|\tilde{w}_n - x\| = \text{diam} K.$$

Proof. The proofs of *(i)* and *(ii)* are easy. Let us first show *(iii)*.

Since $\text{Fix}(\tilde{T})$ is closed, it is enough to prove that for any $\tilde{x}, \tilde{y}$ in $\text{Fix}(\tilde{T})$, there exists $\tilde{z} \in \text{Fix}(\tilde{T})$ such that

$$\|\tilde{x} - \tilde{z}\| = \|\tilde{y} - \tilde{z}\| = \frac{1}{2}\|\tilde{x} - \tilde{y}\| \qquad (3.3)$$

holds. Put $\tilde{x} = \widetilde{(x_n)}$ and $\tilde{y} = \widetilde{(y_n)}$. Let $n \in \mathbb{N}$ be fixed, and put

$$\delta_n = \max(\|x_n - T(x_n)\|, \|y_n - T(y_n)\|),$$

$$\epsilon_n = \min(\frac{1}{2}, \sqrt{\delta_n}),$$

$$d_n = \frac{1}{2}\|x_n - y_n\|, \text{ and}$$

$$\eta_n = \frac{(1 - \epsilon_n)}{\epsilon_n} \delta_n.$$

Consider

$$K_n = \{z \in K : \max(\|x_n - z\|, \|y_n - z\|) \le d_n + \eta_n\}.$$

Clearly, $\frac{x_n + y_n}{2} \in K_n$. Therefore K_n is a nonempty, closed, convex subset of K. Define a mapping $T_n(z) = (1 - \epsilon_n)T(z) + \epsilon_n \frac{(x_n + y_n)}{2}$ for every $z \in K$; we will now show that K_n is invariant under T_n. Indeed, let $z \in K_n$, so

$$x_n - T_n(z) = (1 - \epsilon_n)(x_n - T(z)) + \epsilon_n \frac{(x_n - y_n)}{2}.$$

We then have

$$\|x_n - T_n(z)\| \le (1 - \epsilon_n)(\|x_n - T(x_n)\| + \|T(x_n) - T(z)\|)$$
$$+ \frac{1}{2}\epsilon_n \|x_n - y_n\|$$
$$\le (1 - \epsilon_n)(\delta_n + d_n + \eta_n) + \epsilon_n d_n$$
$$= d_n + (1 - \epsilon_n)\delta_n + (1 - \epsilon_n)\eta_n$$
$$= d_n + \epsilon_n \eta_n + (1 - \epsilon_n)\eta_n = d_n + \eta_n;$$

similarly,

$$\|y_n - T_n(z)\| \le d_n + \eta_n$$

holds. On the other hand, it is easy to see that T_n is a strict contraction. Hence let z_n be the unique fixed point of T_n in K_n. Then

$$\|x_n - z_n\| \le d_n + \eta_n,$$

$$\|y_n - z_n\| \le d_n + \eta_n, \text{ and}$$

$$\|z_n - T(z_n)\| \le \epsilon_n \operatorname{diam} K$$

hold for all n. Since $\tilde{x}$ and $\tilde{y}$ are in $\operatorname{Fix}(\tilde{T})$, one can deduce that

$$\lim_{\mathcal{U}} \delta_n = 0, \tag{3.4}$$

from which we clearly have $\lim_{\mathcal{U}} \epsilon_n = 0$. From (3.4) one can also conclude that $\lim_{\mathcal{U}} \eta_n = 0$. Therefore $\tilde{z} = \widetilde{(z_n)}$ satisfies (3.3), and clearly $\tilde{z} \in \operatorname{Fix}(\tilde{T})$.

III. Basic Results in Ultraproduct Language

Let us now prove statement *(iv)*. Let $\widetilde{(w_n)}$ be a fixed a.f.p.s. for $\widetilde{T}$ in $\widetilde{K}$ and $x \in K$. It is enough to show that $\operatorname{diam} K$ is the only cluster point of $(\|\widetilde{w}_n - x\|)$. Without loss of generality, we can assume that $\lim_{n \to \infty} \|\widetilde{w}_n - x\| = d$. Write $\delta_n = \|\widetilde{w}_n - \widetilde{T}\widetilde{w}_n\|$ and let (ϵ_k) satisfy $\lim_{k \to \infty} \epsilon_k = 0$. If we put $\widetilde{w}_n = \widetilde{(w_m^n)}$ then

$$A_k = \{m \in \mathbf{N} : \|w_m^n - x\| \le d + 2\epsilon_k \text{ and } \|w_m^n - T(w_m^n)\| \le \epsilon_k + \delta_n\}$$

is an element of $\mathcal{U}$ for big enough n. Since $\mathcal{U}$ is non-trivial, we can find strictly increasing sequences of integers $(n(k))$ and $(m(k))$ such that

$$\|w_{m(k)}^{n(k)} - x\| \le d + 2\epsilon_k$$

and

$$\|w_{m(k)}^{n(k)} - T(w_{m(k)}^{n(k)})\| \le \epsilon_k + \delta_{n(k)}$$

hold for every $k \in \mathbf{N}$. The sequence $(w_{m(k)}^{n(k)})$ is an a.f.p.s. for T, and therefore by Proposition 2.9 we have

$$\operatorname{diam} K \le d,$$

which completes the proof. ∎

An interesting application of *(iv)* is the following.

Corollary 3.2. *Let* $\widetilde{W}$ *be any non-empty closed convex subset of* $\widetilde{K}$ *which is invariant under* $\widetilde{T}$. *Then*

$$\sup\{\|x - \widetilde{w}\| : \widetilde{w} \in \widetilde{W}\} = \operatorname{diam} K \qquad (3.5)$$

holds for any $x \in K$.

Proof. By a classical argument, there exists an a.f.p.s. $(\widetilde{w}_n)$ for $\widetilde{T}$ in $\widetilde{W}$. By Theorem 3.1 *(iv)*, we deduce that

$$\lim_{n \to \infty} \|\widetilde{w}_n - x\| = \operatorname{diam} K$$

for any $x \in K$. This clearly implies the desired conclusion (3.5). ∎

IV. SOME FIXED POINT THEOREMS

The main goal of this section is to describe how ultraproduct techniques can be utilized to obtain new fixed point theorems. This elegant technique was originally used by Maurey [149], right after Alspach's counterexample appeared [2]. Maurey proved that reflexive subspaces of L^1 have f.p.p., despite the fact that L^1 does not have f.p.p.

First let us prove some technical results concerning Banach spaces with Schauder bases which will be of interest later in this section. Let X be a Banach space with a Schauder basis (e_n). Let (x_n) be a sequence which converges weakly to 0 in X. Using Proposition 0.14 of Chapter 0, one can find a subsequence (x'_n) of (x_n) and a sequence of natural projections (P_{F_n}), where (F_n) is a sequence of disjoint successive intervals of $\mathbb{N}$ such that

$$\lim_{n \to \infty} \|P_{F_n}(x'_n) - x'_n\| = 0. \tag{4.1}$$

If we denote P_{F_n} by P_n, we can use the properties of (F_n) to deduce the following.

$$P_n \circ P_m = 0 \text{ if } n \neq m; \tag{4.2}$$

$$\lim_{n \to \infty} \|P_n(x)\| = 0 \text{ for any } x \in X. \tag{4.3}$$

We associate new constants to the Schauder basis as follows:

$$\mu = \sup\{\|u - v\| : u \text{ and } v \text{ are disjoint blocks on } (e_n) \text{ with } \|u + v\| \leq 1\}; \tag{4.4}$$

$$c_1 = \sup\{\|I - P_n\| : P_n \text{ is the natural projection on the segment } (1, n) \text{ for any } n \geq 0\}; \tag{4.5}$$

$$c_2 = \sup\{\|I - P_F\| : F \text{ is any seqment in } \mathbb{N}\}. \tag{4.6}$$

Also recall the basis constant of (e_n):

$$c = \sup\{\|P_n\| : P_n \text{ is the natural projection on } [1, n] \text{ for any } n \geq 0\}. \tag{4.7}$$

It is easy to prove that the constants μ, c_1, c_2, c are finite.

In the next theorem the connection between the fixed point property and a Schauder basis is given. The main idea is due to Lin [137].

IV. Some Fixed Point Theorems

Theorem 4.1. *Let X be a Banach space with a Schauder basis (e_n). Assume that the constants μ, c_1, c_2, c satisfy*

$$c_1\mu + c + c_2 < 4; \qquad (4.8)$$

then X has f.p.p.

Proof. Assume that X fails to have f.p.p., so there exists a nonempty weakly-compact convex subset C of X and a nonexpansive mapping $T: C \to C$ with $\text{Fix}(T) = \phi$. Let K be a minimal set for T; without loss of generality we can assume that $\text{diam} K = 1$.

Let (x_n) be an a.f.p.s. in K for T. Since K is weakly compact, we can assume that (x_n) is weakly convergent. Also, the fixed point problem is invariant under translation, so we can assume that the weak limit of (x_n) is 0. Let (P_n) be the sequence of natural projections satisfying (4.1), (4.2), and (4.3). Finally, we can also assume that

$$\lim_{n \to \infty} \|x_{n+1} - x_n\| = 1 \qquad (4.9)$$

holds, from Proposition 2.9.

Let $(X)_{\mathcal{U}}$ be an ultrapower of X, where $\mathcal{U}$ is a nontrivial ultrafilter on $\mathbf{N}$, and let $\widetilde{K}$ and $\widetilde{T}$ be as defined before. Consider

$$\widetilde{x} = \widetilde{(x_n)} \text{ and } \widetilde{y} = \widetilde{(x_{n+1})} \text{ in } \widetilde{K}.$$

Clearly $\widetilde{x}$ and $\widetilde{y}$ are in $\text{Fix}(\widetilde{T})$. Define the operators

$$\widetilde{P} = (P_n)_{\mathcal{U}} \text{ and } \widetilde{Q} = (Q_n)_{\mathcal{U}},$$

where $Q_n = I - \hat{P}_n$ and $\hat{P}_n$ is the completed projection associated with P_n, i.e. if F_n is the segment on which P_n projects, then $\hat{P}_n$ projects on $[1, \max F_n]$.

Using (4.1), (4.2), and (4.3) together with $\widetilde{P}$ and $\widetilde{Q}$, we obtain

$$\widetilde{P}(\widetilde{x}) = \widetilde{x}, \ \widetilde{Q}(\widetilde{y}) = \widetilde{y}, \ \widetilde{P}(\widetilde{y}) = \widetilde{Q}(\widetilde{x}) = \widetilde{P}(x) = \widetilde{Q}(x) = 0 \qquad (4.10)$$

for any $x \in X$. Moreover, by (4.9) we have

$$\|x + y\| = \|\widetilde{P}\widetilde{x} + \widetilde{Q}\widetilde{y}\| = \lim_{\mathcal{U}} \|P_n(x_n) + Q_n(x_{n+1})\|.$$

But

$$\|P_n(x_n) + Q_n(x_{n+1})\| \le \mu \|P_n(x_n) - Q_n(x_{n+1})\|;$$

therefore
$$\|\tilde{x} + \tilde{y}\| \le \mu\|\tilde{P}\tilde{x} - \tilde{Q}\tilde{y}\| = \mu\|\tilde{x} - \tilde{y}\| = \mu. \tag{4.11}$$

Using the definitions of $\tilde{P}$ and $\tilde{Q}$, we obtain

$$\|\tilde{P} + \tilde{Q}\| \le c_1, \ \|I - \tilde{P}\| \le c_2, \text{ and } \|I - \tilde{Q}\| \le c,$$

where I is the identity operator of $\tilde{x}$.

Now set

$$\widetilde{W} = \{\tilde{w} \in \tilde{K} : \text{there exists } x \in K \text{ such that } \|\tilde{w} - x\| \le \frac{\mu}{2};$$

$$\|\tilde{w} - \tilde{x}\| \le \frac{1}{2}; \text{ and } \|\tilde{w} - \tilde{y}\| \le \frac{1}{2}\}.$$

$\widetilde{W}$ is a closed convex subset of $\tilde{K}$. Using (4.11), we deduce that $\frac{\tilde{x}+\tilde{y}}{2} \in \widetilde{W}$, since $0 \in K$. It is easy to see that $\widetilde{W}$ is invariant under $\tilde{T}$, since $\tilde{T}x = Tx \in K$ whenever $x \in K$ and $\tilde{x}$ and $\tilde{y}$ are in $\text{Fix}(\tilde{T})$.

Let $\tilde{w} \in \widetilde{W}$ and x be in K such that $\|\tilde{w} - x\| \le \frac{\mu}{2}$ holds. Then

$$2\tilde{w} = (\tilde{P} + \tilde{Q})\tilde{w} + (I - \tilde{P})\tilde{w} + (I - \tilde{Q})\tilde{w}$$
$$= (\tilde{P} + \tilde{Q})(\tilde{w} - x) + (I - \tilde{P})(\tilde{w} - \tilde{x}) + (I - \tilde{Q})(\tilde{w} - \tilde{y}),$$

so that

$$2\|\tilde{w}\| \le \|\tilde{P} + \tilde{Q}\| \|\tilde{w} - x\| + \|I - \tilde{P}\| \|\tilde{w} - \tilde{x}\| + \|I - \tilde{Q}\| \|\tilde{w} - \tilde{y}\|$$
$$\le c_1\frac{\mu}{2} + c_2\frac{1}{2} + c\frac{1}{2}.$$

Hence
$$\sup\{\|\tilde{w}\| : \tilde{w} \in \widetilde{W}\} \le \frac{(\mu c_1 + c_2 + c)}{4}.$$

Now using corollary 3.2 of Chapter 3, we deduce that

$$1 = \text{diam}K \le \frac{1}{4}(\mu c_1 + c_2 + c).$$

This is a contradiction, so the proof is complete. ∎

As an application of Theorem 4.1, we obtain the following result [99].

77

IV. Some Fixed Point Theorems

Theorem 4.2. *The James quasi-reflexive space J has the fixed point property.*

Proof. Let us evaluate the constants c, c_1, c_2, μ associated with the natural Schauder basis of J. In section II of Chapter 2 we proved that $\mu \leq \sqrt{2}$ and $c_2 \leq \sqrt{2}$ (one can have equalities), and the constants c and c_1 are equal to 1 since the basis of J is bimonotone. Clearly $\sqrt{2} + \sqrt{2} + 1 < 4$ holds, so by Theorem 4.1 we deduce that J has f.p.p. ∎

Remark 4.3. Since some authors call J_0 (introduced in section VI of Chapter 2) the James quasi-reflexive space. One may ask whether J_0 has f.p.p.; Theorem 4.1 does not give an answer since the constants associated with the basis of J_0 are big enough not to satisfy (4.8). However, in [104] it is proven (not using ultraproduct techniques) that weakly compact, convex subsets of J_0 have normal structure, so that by Theorem 2.7 J_0 has f.p.p.

The second application of Theorem 4.1 concerns the Banach spaces X_β defined in section II as $X_\beta = (\ell_2, \|\cdot\|_\beta)$, where

$$\|x\|_\beta = \max(\|x\|_{\ell_2}, \beta\|x\|_{\ell_\infty}) \text{ for any } x \in \ell_2.$$

We have seen that X_β has f.p.p. for $\beta \leq 2$. Clearly the canonical basis of ℓ_2 is also an unconditional Schauder basis for X_β, with the constant of unconditionality equal to 1. Therefore the constants c, c_1, c_2, μ are all 1 for this basis. So Theorem 4.1 allows us to deduce that X_β has the fixed point property for any $\beta > 0$.

More generally, Lin [139] proved the following.

Theorem 4.4 *Let X be a Banach space with an unconditional Schauder basis. Assume that the constant of unconditionality is 1. Then X has the fixed point property.*

Proof. Since the constant of unconditionality of X is 1, we deduce that all the constants c, c_1, c_2, μ are equal to 1. Therefore, by using Theorem 4.1 we deduce that X has f.p.p.

In the next theorem, we consider Banach spaces with unconditional Schauder Bases for which the constant of unconditionality is not 1. (See [139].)

78

Theorem 4.5. *Let X be a Banach space with an unconditional Schauder basis (e_n), and suppose that X fails to have f.p.p. Then the constant of unconditionality λ of (e_n) satisfies the inequality*

$$\lambda^2 + 3\lambda - 6 \geq 0, \text{ or equivalently}$$

$$\lambda \geq \frac{\sqrt{33} - 3}{2}.$$

Proof. Since X fails to have f.p.p., there exists a nonempty weakly compact convex subset K which is a minimal set for a nonexpansive mapping T, with $\operatorname{diam} K = 1$.

As in the proof of Theorem 4.1, we can find an a.f.p.s. (x_n) which is weakly convergent to 0 and a sequence of natural projections (P_n) such that (4.1), (4.2), (4.3), (4.9) hold. Let $\mathcal{U}$ be a nontrivial ultrafilter on $\mathbb{N}$, and consider the ultrapower $(X)_\mathcal{U}$ of X. Define $\tilde{K}$ and $\tilde{T}$ as usual. Set $\tilde{x} = \widetilde{(x_n)}$, $\tilde{y} = \widetilde{(x_{n+1})}$, and $\tilde{P} = (P_n)_\mathcal{U}$, as done in the proof of Theorem 4.1, and let $\tilde{Q} = (P_{n+1})_\mathcal{U}$. We have

$$\tilde{P}\tilde{x} = \tilde{x}, \tilde{Q}\tilde{y} = \tilde{y}, \tilde{P}(\tilde{y}) = \tilde{Q}(\tilde{x}) = \tilde{P}(x) = \tilde{Q}(x) = 0$$

for any $x \in X$. One can also obtain $\|\tilde{x} + \tilde{y}\| \leq \lambda$.

Consider

$$\widetilde{W} = \{\tilde{w} \in \tilde{K} : \text{there exists } x \in K \text{ such that}$$

$$\|\tilde{w} - x\| \leq \frac{\lambda}{2} \text{ and } \max(\|\tilde{x} - \tilde{w}\|, \|\tilde{y} - \tilde{w}\|) \leq \frac{1}{2}\}.$$

Clearly $\widetilde{W}$ is a closed, convex subset of $\tilde{K}$ and is invariant under $\tilde{T}$. And since $\frac{\tilde{x}+\tilde{y}}{2} \in \widetilde{W}$, we deduce from Corollary 3.2 of Chapter 3 that

$$\sup\{\|\tilde{w}\| : \tilde{w} \in \widetilde{W}\} = 1.$$

To simplify the calculations we may assume that $\widetilde{W}$ contains a $\tilde{w}$ with $\|\tilde{w}\| = 1$. Let $x \in K$ be such that $\|\tilde{w} - x\| \leq \frac{\lambda}{2}$, and let $\tilde{f} \in \tilde{X}^*$ with $\|\tilde{f}\| = 1$ and $\tilde{f}(\tilde{w}) = 1$. Hence

$$1 - \tilde{f}(\tilde{x}) = \tilde{f}(\tilde{w} - \tilde{x}) \leq \|\tilde{w} - \tilde{x}\| \leq \frac{1}{2},$$

or $\tilde{f}(\tilde{x}) \geq \frac{1}{2}$. Similarly we have $\tilde{f}(\tilde{y}) \geq \frac{1}{2}$ and $\tilde{f}(x) \geq 1 - \frac{\lambda}{2}$. Put $\alpha = \tilde{f}((I - \tilde{P} - \tilde{Q})(\tilde{w}))$; then

$$1 - \alpha = \tilde{f}(\tilde{w}) - \tilde{f}((I - \tilde{P} - \tilde{Q})(\tilde{w})) = \tilde{f}((\tilde{P} + \tilde{Q})(\tilde{w})).$$

Therefore either $\tilde{f}(\tilde{P}(\tilde{w})) \leq \frac{1-\alpha}{2}$ or $\tilde{f}(\tilde{Q}(\tilde{w})) \leq \frac{1-\alpha}{2}$. Suppose the former. Since $I - 2\tilde{P}$ and $2\tilde{P} + 2\tilde{Q} - I$ are reflections, then $\|I - 2\tilde{P}\| \leq \lambda$ and $\|2\tilde{P} + 2\tilde{Q} - I\| \leq \lambda$. Hence

$$\begin{aligned}
2 - 2\alpha - \frac{\lambda}{2} &\leq 2\tilde{f}[(\tilde{P} + \tilde{Q})(\tilde{w})] - \tilde{f}(\tilde{w} - x) \\
&= 2\tilde{f}[(\tilde{P} + \tilde{Q})(\tilde{w} - x)] - \tilde{f}(\tilde{w} - x) \\
&= \tilde{f}[(2\tilde{P} + 2\tilde{Q} - I)(\tilde{w} - x)] \\
&\leq \|(2\tilde{P} + 2\tilde{Q} - I)(\tilde{w} - x)\| \\
&\leq \|2\tilde{P} + 2\tilde{Q} - I\| \, \|\tilde{w} - x\| \leq \frac{\lambda}{2} \cdot \lambda = \frac{\lambda^2}{2}.
\end{aligned}$$

On the other hand, we have

$$\begin{aligned}
\frac{1}{2} + \alpha &= \frac{1}{2} + 1 - (1 - \alpha) \\
&\leq \tilde{f}(\tilde{x}) + \tilde{f}(\tilde{w}) - 2\tilde{f}(\tilde{P}\tilde{w}) \\
&= \tilde{f}(\tilde{w} - \tilde{x}) + 2\tilde{f}(\tilde{x}) - 2\tilde{f}(\tilde{P}\tilde{w}) \\
&= \tilde{f}(\tilde{w} - \tilde{x}) + \tilde{f}(2\tilde{P}(\tilde{x} - \tilde{w})) \\
&= \tilde{f}[(I - 2\tilde{P})(\tilde{w} - \tilde{x})] \\
&\leq \lambda \|\tilde{w} - \tilde{x}\| \leq \frac{\lambda}{2}.
\end{aligned}$$

Therefore $2 - 2\alpha - \frac{\lambda}{2} + 2(\frac{1}{2} + \alpha) \leq \frac{\lambda^2}{2} + \lambda$, which gives the desired inequality. ∎

Remark 4.6. One might think that the value $\lambda_0 = \frac{\sqrt{33}-3}{2}$ is sharp, since we used the Hahn-Banach theorem in the proof. However, in [105], J. B. Baillon gave another proof of Theorem 4.5 using the triangle inequality instead of the Hahn-Banach theorem. For the sake of completeness, we give this proof also.

As in the proof just given, let K, T, and (x_n) satisfy (4.1), (4.2), (4.3), (4.9) for a given sequence of projections (P_n). We also consider

$\tilde{x}$, $\tilde{y}$, $\tilde{P}$, and $\tilde{Q}$ as in the proof above. Denote the set of quasi-midpoints of $\tilde{x}$ and $\tilde{y}$ in $\tilde{K}$ by $\frac{1}{2}[\tilde{x}, \tilde{y}]$, i.e.

$$\frac{1}{2}[\tilde{x}, \tilde{y}] = \{\tilde{z} \in \tilde{K} : \|\tilde{x} - \tilde{z}\| = \|\tilde{y} - \tilde{z}\| = \frac{1}{2}\|\tilde{x} - \tilde{y}\|\}.$$

Put

$$\mu = d(\frac{1}{2}[\tilde{x}, \tilde{y}], K) = \inf\{\|\tilde{w} - x\| : \tilde{w} \in \frac{1}{2}[\tilde{x}, \tilde{y}] \text{ and } x \in K\}.$$

Let $\epsilon > 0$ be fixed and set

$$\tilde{K}_\epsilon = \{\tilde{w} \in \frac{1}{2}[\tilde{x}, \tilde{y}] : d(\tilde{w}, K) \le \mu + \epsilon\}.$$

Hence $\tilde{K}_\epsilon$ is a nonempty convex subset of $\tilde{K}$ which is invariant under $\tilde{T}$. Let $\tilde{w} \in \tilde{K}_\epsilon$ and set $\tilde{u} = (\tilde{P} + \tilde{Q})(\tilde{w})$ and $\tilde{r} = \tilde{w} - \tilde{u}$. Then

$$2\tilde{u} = (\tilde{u} + \tilde{r} - x) + (\tilde{u} - \tilde{r} + x)$$

for any $x \in K$. Using the definitions of $\tilde{P}$ and $\tilde{Q}$, we deduce

$$\|\tilde{u} - \tilde{r} + x\| \le \lambda \|\tilde{u} + \tilde{r} - x\|.$$

Hence

$$2\|\tilde{u}\| \le (\lambda + 1)\|\tilde{u} + \tilde{r} - x\| = (\lambda + 1)\|\tilde{w} - x\|$$

and therefore

$$2\|\tilde{u}\| \le (\lambda + 1)(\mu + \epsilon). \tag{4.12}$$

On the other hand, we have

$$\tilde{w} + \tilde{y} - 2\tilde{P}(\tilde{w}) = \tilde{P}(\tilde{w}) + \tilde{Q}(\tilde{w}) + \tilde{r} - 2\tilde{P}(\tilde{w}) + \tilde{y}$$
$$= \tilde{Q}(\tilde{w}) + \tilde{r} - \tilde{P}(\tilde{w}) + \tilde{y}.$$

Hence

$$\|\tilde{w} + \tilde{y} - 2\tilde{P}(\tilde{w})\| \le \lambda \|\tilde{Q}(\tilde{w}) + \tilde{r} - \tilde{y} + \tilde{P}(\tilde{w})\|$$
$$= \lambda \|\tilde{w} - \tilde{y}\| = \frac{\lambda}{2} \tag{4.13}$$

since $\tilde{w} \in \frac{1}{2}[\tilde{x}, \tilde{y}]$ and $\|\tilde{x} - \tilde{y}\| = \operatorname{diam} K = 1$. Similarly, we have

$$\|\tilde{w} + \tilde{x} - 2\tilde{Q}(\tilde{w})\| \le \frac{\lambda}{2}. \tag{4.14}$$

IV. Some Fixed Point Theorems

Now (4.13) and (4.14) imply

$$\|\widetilde{w} + \frac{\widetilde{y} + \widetilde{x}}{2} - \widetilde{P}(\widetilde{w}) - \widetilde{Q}(\widetilde{w})\| \le \frac{\lambda}{2}. \qquad (4.15)$$

We then have, by (4.14) and (4.15),

$$\begin{aligned}
\|\widetilde{w} + \frac{\widetilde{x} + \widetilde{y}}{2}\| &\le \|\widetilde{w} - \widetilde{r}\| + \|\widetilde{r} + \frac{\widetilde{x} + \widetilde{y}}{2}\| \\
&\le \frac{\lambda + 1}{2}(\mu + \epsilon) + \frac{\lambda}{2}.
\end{aligned} \qquad (4.16)$$

But

$$\begin{aligned}
\|\widetilde{w} + \frac{\widetilde{x} + \widetilde{y}}{2}\| &= 2\|\widetilde{w} - \frac{1}{2}(\widetilde{w} - \frac{\widetilde{x} + \widetilde{y}}{2})\| \\
&\ge 2\|\widetilde{w}\| - \frac{1}{2}
\end{aligned}$$

since $\|\widetilde{w} - \frac{\widetilde{x}+\widetilde{y}}{2}\| \le \frac{1}{2}$. Therefore by (4.16) we have

$$2\|\widetilde{w}\| \le \frac{\lambda + 1}{2}(\mu + \epsilon) + \frac{\lambda}{2} + \frac{1}{2}$$

for any $\widetilde{w} \in \widetilde{K}_\epsilon$. Using Corollary 3.2 of Chapter 3, we deduce that

$$2\sup\{\|\widetilde{w}\| : \widetilde{w} \in \widetilde{K}_\epsilon\} = 2 \le \frac{\lambda + 1}{2}(\mu + \epsilon) + \frac{\lambda}{2} + \frac{1}{2}.$$

On the other hand, it is easy to see that

$$\mu \le \frac{\lambda}{2}.$$

So

$$2 \le \frac{\lambda + 1}{2}(\frac{\lambda}{2} + \epsilon) + \frac{\lambda}{2} + \frac{1}{2}.$$

And since ϵ was arbitrary, we obtain the desired inequality of Theorem 4.5. ∎

Another consequence of Theorem 4.1 concerns Banach spaces with an unconditional basis for which the unconditional basis constant is 1. We say that an unconditional basic sequence (e_n) is *unconditionally monotone* if $c = \sup\{\|P_F\| : F \text{ is any subset of } \mathbf{N}\}$ is equal to 1. One can easily deduce from Theorem 4.1 that any Banach space with an unconditional Schauder basis which is unconditionally monotone and for which the constant of unconditionality satisfies $\lambda < 2$, has the fixed point property. The next theorem deals with the extreme case $c = 1$ and $\lambda = 2$ (see [103]).

Theorem 4.7. *Let X be a Banach space which has an unconditional basis which is unconditionally monotone. Assume that X has the alternate Banach-Saks property. Then X has f.p.p.*

Proof. Suppose that X fails to have f.p.p. Then, as in the proof of Theorem 4.1, we can find a nonempty weakly compact convex subset K which is a minimal set for a nonexpansive mapping T; an a.f.p.s. (x_n) which converges weakly to 0; and a sequence of natural projections (P_n) such that (4.1), (4.2), (4.3), and (4.9) hold. As usual we can assume that $\operatorname{diam} K = 1$.

Since any sequence without a convergent subsequence has a "good" subsequence, by Theorem 2.5 of Chapter 2, we can assume without loss of generality that (x_n) is a "good" sequence, i.e.

$$\lim_{\substack{n_1 < n_2 < \cdots < n_k \\ n_1 \to \infty}} \left\| \sum_{i=1}^{k} c_i x_{n_i} \right\|$$

exists for any finite scalars $c_1, c_2, \ldots, c_k$. Let $\mathcal{U}$ be a nontrivial ultrafilter on $\mathbf{N}$, and define $\widetilde{K}$ and $\widetilde{T}$ in $(X)_{\mathcal{U}}$ as before. Let $k \in \mathbf{N}$ be fixed. To each integer n we associate the integers $n_1, n_2, \ldots, n_k$ such that $n_1 < n_2 < \cdots < n_k$ holds. Define $\widetilde{x}_i = \widetilde{(x_{n_i})}$ and $\widetilde{P}_i = (P_{n_i})_{\mathcal{U}}$ for all $i = 1, 2, \ldots, k$. Then we have

(1) $\widetilde{P}_i \widetilde{P}_j = 0$ for $i \neq j$;

(2) $\widetilde{P}_i(\widetilde{x}_i) = \widetilde{x}_i$ for any $i = 1, \ldots, k$;

(3) $\widetilde{P}_i(x) = 0$ for any $i = 1, \ldots, k$ and any $x \in X$.

Let $(\alpha_i)_{i \leq k}$ satisfy $\alpha_i \geq 0$ and $\sum_{i=1}^{k} \alpha_i = 1$. For each $0 < \beta < 1$ set

$$\widetilde{W}_\beta = \{\widetilde{w} \in \widetilde{K} : \text{there exists } x \in K \text{ such that } \|\widetilde{w} - x\| \leq \beta$$
$$\text{and } \|\widetilde{w} - \widetilde{x}_i\| \leq 1 - \alpha_i \text{ for any } i = 1, \ldots, k\}.$$

Obviously, $\widetilde{W}_\beta$ is a closed, convex subset of $\widetilde{K}$ and is invariant under $\widetilde{T}$. Suppose that $\widetilde{W}_\beta$ is not empty and let $\widetilde{w} \in \widetilde{W}_\beta$; then there exists

IV. Some Fixed Point Theorems

$x \in K$ such that $\|\widetilde{w} - x\| \leq \beta$. Then we have

$$k\widetilde{w} = \sum_{i=1}^{k}(I - \widetilde{P}_i)(\widetilde{w}) + \sum_{i=1}^{k}\widetilde{P}_i(\widetilde{w})$$

$$= \sum_{i=1}^{k}(I - \widetilde{P}_i)(\widetilde{w} - \widetilde{x}_i) + \sum_{i=1}^{k}\widetilde{P}_i(\widetilde{w} - x).$$

By taking norms, and keeping in mind that the basis is unconditionally monotone, we obtain

$$k\|\widetilde{w}\| \leq \sum_{i=1}^{k}(1 - \alpha_i) + \beta = k - 1 + \beta.$$

Thus $\sup\{\|\widetilde{w}\| : \widetilde{w} \in \widetilde{W}_\beta\} \leq \frac{k-1+\beta}{k} < 1$, and this contradiction implies that $\widetilde{W}_\beta = \phi$. Therefore we have $\|\sum_{i=1}^{k}\alpha_i\widetilde{x}_i\|_{(X)_{\mathcal{U}}} \geq \beta$, since $\|\widetilde{x}_i - \sum_{j=1}^{k}\alpha_j\widetilde{x}_j\| \leq 1 - \alpha_i$. Hence

$$\|\sum_{i=1}^{k}\alpha_i\widetilde{x}_i\| = 1, \qquad (4.17)$$

because β was arbitrary in $)0,1($.

On the other hand, since (x_n) is a good sequence we have

$$\|\sum_{i=1}^{k}a_i\widetilde{x}_i\|_{(X)_{\mathcal{U}}} = \lim_{n,\mathcal{U}}\|\sum_{i=1}^{k}a_ix_{n_i}\|$$

$$= \|\sum_{i=1}^{k}a_ie_i\|_F, \qquad (4.18)$$

where (e_n) is the fundamental basis of the spreading model F generated by (x_n).

Let us prove that (e_n) is equivalent to the canonical basis of ℓ_1. Indeed, since

$$\|\sum_{i=1}^{k}a_i\widetilde{x}_i\| = \|\sum_{i=1}^{k}a_i\widetilde{P}(\widetilde{x}_i)\|$$

we obtain

$$\left\|\sum_{i=1}^{k} a_i \tilde{x}_i\right\| \leq \lambda \left\|\sum_{i=1}^{k} a_i \epsilon_i \tilde{x}_i\right\|$$

for any sequence of signs (ϵ_i) (i.e. $\epsilon_i = \pm 1$). It then follows, using (4.17), that

$$\frac{1}{\lambda}\sum_{i=1}^{k}|a_i| \leq \left\|\sum_{i=1}^{k} a_i \tilde{x}_i\right\| \leq \lambda \sum_{i=1}^{k}|a_i|. \tag{4.19}$$

From (4.18) we deduce

$$\frac{1}{\lambda}\sum_{i=1}^{k}|a_i| \leq \left\|\sum_{i=1}^{k} a_i e_i\right\|_F \leq \lambda \sum_{i=1}^{k}|a_i|.$$

Since k was an arbitrary nonnegative integer, we can conclude that (e_i) is equivalent to the canonical basis of ℓ_1. But by Theorem 4.8 of Chapter 2, this implies that X cannot have the alternate Banach-Saks property, which is a contradiction. ∎

Remarks 4.8. (1) In the above proof a sharper estimate than (4.19) is possible. Let $(\alpha_i)_{i\leq k}$ be any scalars and set $\alpha_i^+ = \max(\alpha_i, 0)$ and $\alpha_i^- = \max(-\alpha_i, 0)$ for each i. Then $\sum_{i=1}^{k} \alpha_i^+ \tilde{x}_i \in \tilde{K}$ and $\sum_{i=1}^{k} \alpha_i^- \tilde{x}_i \in \tilde{K}$, given that $\max(\sum_{i=1}^{k} \alpha_i^+, \sum_{i=1}^{k} \alpha_i^-) = 1$. So $\|\sum_{i=1}^{k} \alpha_i^+ \tilde{x}_i - \sum_{i=1}^{k} \alpha_i^- \tilde{x}_i\| \leq 1$ holds, since one of α_i^+ or α_i^- is equal to 0 for all i. From the properties of $(\tilde{P}_i)$ we can deduce that

$$\left\|\sum \alpha_i^+ \tilde{x}_i\right\| \leq \left\|\sum \alpha_i^+ \tilde{x}_i - \sum \tilde{\alpha}_i^- \tilde{x}_i\right\| \leq 1$$

and

$$\left\|\sum \alpha_i^- \tilde{x}_i\right\| \leq \left\|\sum \alpha_i^+ \tilde{x}_i - \sum \tilde{\alpha}_i^- \tilde{x}_i\right\| \leq 1.$$

By (4.17), we have

$$\left\|\sum \alpha_i^+ \tilde{x}_i\right\| = \sum \alpha_i^+ \text{ and } \left\|\sum \alpha_i^- \tilde{x}_i\right\| = \sum \alpha_i^-.$$

Hence $\max(\sum \alpha_i^+, \sum \alpha_i^-) = 1 \leq \|\sum \alpha_i^+ \tilde{x}_i - \sum \alpha_i^- \tilde{x}_i\| \leq 1$ holds. Finally, since $\|\sum \alpha_i \tilde{x}_i\| = \|\sum \alpha_i^+ \tilde{x}_i - \sum \alpha_i^- \tilde{x}_i\|$, we obtain

$$\left\|\sum \alpha_i \tilde{x}_i\right\| = \max\left(\sum \alpha_i^+, \sum \alpha_i^-\right)$$

for any scalars (α_i).

(2) One may obtain a similar conclusion even if X does not have an unconditional basis. Indeed, let $K, T, \widetilde{K}$, and $\widetilde{T}$ be as in the proof of Theorem 4.7. To each $\widetilde{x} = \widetilde{(x_n)}$ associate the set

$$C_{\widetilde{x}} = \{\widetilde{x}_\beta = (x_{\widetilde{\beta(n)}}) : \beta : \mathsf{N} \to \mathsf{N} \text{ is strictly increasing}\}.$$

Note that if (x_n) is an a.f.p.s. for T, then $C_{\widetilde{x}} \subset \text{Fix}(\widetilde{T})$, where $\widetilde{x} = \widetilde{(x_n)}$.

Assume that there exists an a.f.p.s. (x_n) for T which is weakly convergent to 0, but that $0 \notin \overline{\text{conv}} C_{\widetilde{x}}$, where $\widetilde{x} = \widetilde{(x_n)}$. Then the Hahn-Banach theorem implies that there exists $\delta > 0$ such that

$$\|\widetilde{y}\| \geq \delta \text{ for any } \widetilde{y} \in \overline{\text{conv}} C_{\widetilde{x}}. \tag{4.20}$$

Without loss of generality, we can assume that (x_n) is a "good sequence," i.e. that

$$\lim_{\substack{n_1 < n_2 < \cdots < n_k \\ n_1 \to \infty}} \left\| \sum_{i=1}^{k} c_i x_{n_i} \right\|$$

exists for any fixed $k \in \mathsf{N}$ and scalars (c_i). Let F be the spreading model generated by (x_n), with fundamental basis (e_n). Then by (4.20) we obtain

$$\left\| \sum_{i=1}^{k} \alpha_i e_i \right\|_F \geq \delta \sum_{i=1}^{k} \alpha_i \tag{4.21}$$

for any positive sequence of scalars $(\alpha_i)_{i \leq k}$.

Suppose that (e_n) is not equivalent to the natural basis of ℓ_1; then (e_n) converges to 0, since (x_n) converges weakly to 0 (see [82]). Therefore (e_n) is an unconditional basis (see [18]), and (4.21) implies that (e_n) is equivalent to the natural basis of ℓ_1. In other words, (e_n) is obliged to be equivalent to the canonical basis of ℓ_1, and therefore X does not have the alternate Banach-Saks property.

Let us give an example of mappings for which the condition on $C_{\widetilde{x}}$ is satisfied.

Definition 4.9. A mapping T is said to be *of convex type* if $\text{Fix}(\widetilde{T})$ is convex.

It is clear that if T is of convex type, then $\text{conv}C_{\tilde{x}} \subset \text{Fix}(\tilde{T})$. Therefore $d(x, \text{conv}C_{\tilde{x}}) = \text{diam}K$ for any $x \in K$. Since the fixed point set of a nonexpansive mapping in a strictly convex Banach space is a convex set, we deduce that any nonexpansive mapping defined in a Banach space X is of convex type if $(X)_u$ is strictly convex.

Recall that $(X)_u$ is strictly convex if any only if X is uniformly convex. In this case Bruck [32] has proven that nonexpansive mappings in uniformly convex Banach spaces are of type (γ).

Definition 4.10. Let K be a bounded, closed, convex subset of X and let $T : K \to X$ be a map. We say that T is *of type* (γ) if and only if there exists a $\gamma \in \Gamma$ such that

$$\gamma(\|T(\frac{x+y}{2}) - \frac{T(x)+T(y)}{2}\|) \leq \|x-y\| - \|T(x) - T(y)\| \quad (4.22)$$

for any $x, y \in K$, where

$$\Gamma = \{\gamma : \mathbb{R}_+ \to \mathbb{R}_+ : \gamma(0) = 0, \text{ and } \gamma$$

is a strictly increasing continuous function}.

Baillon [10] introduced this class of mappings and proved that in L_p, $p > 1$, any nonexpansive mapping is of type (γ). This result was generalized by Bruck [32] to uniformly convex Banach spaces (see also [106]).

A non-linear mean ergodic theorem for nonexpansive mappings of type (γ) has been proven in [9] (see also [32]). Also, it is clear that if T is of type (γ), then $\tilde{T}$ is also of type (γ). Furthermore, T is of convex type whenever T is of type (γ). A fixed point theorem for mappings of convex type was proven in [105].

As a corollary to Theorem 4.7, we can get the following result, due to Lin [139].

Corollary 4.11. *Let X be a super-reflexive Banach space with an unconditionally monotone basis. Then X has the fixed point property.*

Proof. Since any super-reflexive Banach space has the A.B.S.P., the conclusion is immediate from Theorem 4.7. ∎

Consider the Tzirelson space $\mathcal{T}$ with natural basis (e_n). For any $x = \sum x_n e_n$ in $\mathcal{T}$ set

$$x^+ = \sum x_n^+ e_n \quad \text{and} \quad x^- = \sum x_n^- e_n,$$

where $a^+ = \max(a, 0)$ and $a^- = \max(-a, 0)$. Now define an equivalent norm $|\cdot|$ on $\mathcal{T}$ by

$$|x| = \max(\|x^+\|_{\mathcal{T}}, \|x^-\|_{\mathcal{T}}). \tag{4.23}$$

Lin [139] asked whether $(\mathcal{T}, |\cdot|)$ has f.p.p. The answer to this question is still unknown. Let us remark that the canonical basis of $\mathcal{T}$ is an unconditionally monotone basis. Theorem 4.7 cannot apply in this case, since the fundamental basis of any spreading model of $\mathcal{T}$ is equivalent to ℓ_1 (see [18]), and therefore $\mathcal{T}$ fails to have A.B.S.P. However the dual space $(\mathcal{T}^*, |\cdot|^*)$ of $(\mathcal{T}, |\cdot|)$ does have the fixed point property, since the biorthogonal system (e_n^*) forms an unconditionally monotone basis of $(\mathcal{T}^*, |\cdot|^*)$ and $\mathcal{T}^*$ has A.B.S.P.

The last result of this section is analogous to Theorem 4.1 but dealing with dual Banach spaces. Recall that in section II of this chapter, Proposition 2.11 did the same thing for Proposition 2.9. The following theorem can be proven using Proposition 2.11 and the proof of Theorem 4.1.

Theorem 4.12. *Let X be a Banach space with a shrinking Schauder basis (e_n). Suppose that (e_n) is strongly bimonotone with associated constant $\mu < 2$. Then any weak*-compact convex subset of X^* has the fixed point property.*

V. MAUREY'S THEOREMS

The use of ultraproducts in fixed point theory was initiated by Maurey [148]. In this section, we will discuss some of Maurey's results, such as the fact that any reflexive subspace of L^1 has f.p.p. – this is particularly interesting since Alspach [2] has proven that L^1 fails to have f.p.p.

Since Maurey's proof of this result uses random measures, we begin by recalling certain definitions and results about them. We will work with $L^1(\Omega, \mathcal{A}, P, \mathbf{C})$. Without loss of generality, we will assume that L^1 is separable.

Definition 5.1. Let $(\Omega, \mathcal{A}, P)$ be a probability space and let $P(\mathbf{C})$ be the set of probabilities on $\mathbf{C}$. A *random probability* on Ω is a mapping $\mu : \Omega \to P(\mathbf{C})$ such that

$$w \to \int \varphi(u)\mu_w(du)$$

is measurable for any bounded continuous function φ on $\mathbf{C}$.

We will write $\mu = (\mu_w)_{w \in \Omega}$ for the random measure μ.

Let us recall that the set of bounded continuous functions on $\mathbf{C}$ can be viewed as $\mathcal{C}(\overline{\mathbf{C}})$ (the set of continuous functions on $\overline{\mathbf{C}}$), where by $\overline{\mathbf{C}}$ we mean the Alexandroff compactification of $\mathbf{C}$. It is clear that the set of random probabilities is a subset of the dual space of $L^1(\Omega, \mathcal{A}, P, \mathcal{C}(\overline{\mathbf{C}}))$, where the duality is given by

$$< \mu, f > = \int f(w, u)\mu_w(du)\, dP(w)$$
$$= E\left(\int f(w, u)\mu_w(du) \right),$$

$$(5.1)$$

where E is the integral over $x \in L^1(\Omega, \mathcal{A}, P)$.

The following example shows how a random measure can be generated from a bounded sequence in $L^1(\Omega, \mathcal{A}, P)$.

Example 5.2. Let $x \in L^1(\Omega, \mathcal{A}, P)$ and consider $\mu_x = (\delta_{x(w)})_{w \in \Omega}$, where δ_a is the Dirac measure at a. Clearly μ_x is a random probability on Ω. Now let (x_n) be a bounded sequence in $L^1(\Omega, \mathcal{A}, P)$, with associated probabilities $\mu_n = \mu_{x_n}$. Consider the weak* limit of (μ_n) over a nontrivial ultrafilter $\mathcal{U}$ (on $\mathbf{N}$), in the dual space of $L^1(\Omega, \mathcal{A}, P, \mathcal{C}(\overline{\mathbf{C}}))$. Put $\mu = \text{weak}^* - \lim_{\mathcal{U}} \mu_n$; then we have

$$< \mu, f > = \lim_{n, \mathcal{U}} < \mu_n, f > = \lim_{n, \mathcal{U}} E(f(x_n))$$
$$= \lim_{n, \mathcal{U}} \int f(w, x_n(w))\, dP(w)$$

$$(5.2)$$

for any $f \in L^1(\Omega, \mathcal{A}, P, \mathcal{C}(\overline{\mathbb{C}}))$.

Assume that (x_n) is weakly convergent to x in $L^1(\Omega, \mathcal{A}, P)$. Then (x_n) is equi-integrable (see Definition 2.12 of Chapter 2), so equation (5.2) can be extended to f, which does not increase faster than $|u|$ at ∞. By this we mean that

$$|f(w, u)| \leq M|u| + g(w), \qquad (5.3)$$

where $M > 0$ and $g \in L^1(\Omega, \mathcal{A}, P)$.

Proposition 5.3. *Let (x_n) be as above and let μ be its associated random probability. Then the following hold.*

(1) $\lim_{\mathcal{U}} \|x_n\|_{L^1} = E(\int |u| \mu(du))$;

(2) $\lim_{\mathcal{U}} \|x_n - y\|_{L^1} = E(\int |y(w) - u| \mu(du))$;

(3) $\lim_{n,\mathcal{U}} \lim_{m,\mathcal{U}} \|x_n - x_m\|_{L^1} = E(\int |v - u| \mu(du)\mu(dv))$;

(4) $x(w) = \int u\mu_w(du)$ *for almost all $w \in \Omega$.*

Proof. Statements *(1)*, *(2)*, and *(3)* will follow from an appropriate choice of f which satisfies (5.3). For *(1)* one can take $f(w, u) = |u|$, for *(2)* $f(w, u) = |u - y(w)|$ and for *(3)* $f(w, u) = \int |v - u| \mu_w(dv)$.

To prove statement *(4)*, let $g \in L_\infty(\Omega, \mathcal{A}, P)$. Since (x_n) converges weakly to x, we have

$$\int g(w)x(w)dP(w) = \lim_{n \to \infty} \int g(w)x_n(w)dP(w).$$

Take $f(w, u) = g(w)u$; then f satisfies (5.3) and so we have

$$E(f(x_n)) = \int g(w)x_n(w)dP(w),$$

which implies that

$$\int g(w)x(w)dP(w) = E\left(\int g(w)u\mu_w(du)\right)$$

for any $g \in L_\infty(\Omega, \mathcal{A}, P)$. This completes the proof. $\blacksquare$

Notice that if (x_n) are real functions, then the support of μ_w is in $\mathbb{R}$ for almost every $w \in \Omega$. Furthermore, if $x_n(w) \in [X(w), Y(w)]$

for almost all $w \in \Omega$, then $\mathrm{supp}\mu_w \subset [X(w), Y(w)]$ for almost all $w \in \Omega$. Indeed, if we take

$$f(w, u) = 1 \wedge \mathrm{dist}(u, [X(w), Y(w)])$$

where $a \wedge b$ means $\inf(a, b)$, we have $E(f(x_n)) = 0$ for every $n \in \mathbf{N}$. Then $E(\int f(u)\mu(du)) = 0$, which gives the desired conclusion.

In the following we will discuss how the random probabilities give us information on minimal convex sets for nonexpansive mappings in L^1. Let K be a weakly compact convex subset of $L^1(\Omega, \mathcal{A}, P)$ which is minimal for a nonexpansive mapping T. We will assume that K is not reduced to a single point and, without loss of generality, that $\mathrm{diam} K = 1$.

Proposition 5.4. *Let K and T be as above and let $\mathcal{U}$ be a nontrivial ultrafilter on $\mathbf{N}$. Then the following hold.*

(1) There exists $x_0 \in K$ and a measurable function $G : \Omega \to \mathbf{C}$, with $|G(w)| = 1$ for almost all $w \in \Omega$, such that for every $x \in K$

$$(x - x_0)G \text{ is a real function.} \tag{5.4.}$$

(2) Assume that $K \subset L^1(\Omega, \mathcal{A}, P, \mathbf{R})$. Let $\mathcal{D}$ be a countable index set and (x_n^α) an a.f.p.s. for T in K, for all $\alpha \in \mathcal{D}$. Then there exist two measurable functions U and V on Ω such that

$$\lim_{n,\mathcal{U}} E(|x_n^\alpha - U| \wedge |x_n^\alpha - V|) = 0 \tag{5.5}$$

for any $\alpha \in \mathcal{D}$.

Proof. Let (x_n) be an a.f.p.s. for T in K. Since K is weakly compact, (x_n) is weakly convergent to $x_0 \in K$, with respect to $\mathcal{U}$. By translating appropriately, we may assume that $x_0 = 0$. Using Proposition 2.9, we obtain

$$\lim_{n \to \infty} \|x - x_n\| = \delta(K) = 1$$

for any $x \in K$.

Let μ be the random probability associated with (x_n). Then, by Proposition 5.3, we have

$$E\left(\int |x - u|\mu(du)\right) = \lim_{n,\mathcal{U}} \|x_n - x\| = 1. \tag{5.6}$$

V. Maurey's Theorems

In particular, when $x = 0$ we have

$$E\left(\int |u|\mu(du)\right) = 1.$$

We also have

$$E\left(\int |v - u|\mu(du)\mu(dv)\right) = 1, \tag{5.7}$$

and since (x_n) converges weakly to 0,

$$\int u\mu_w(du) = 0 \quad \text{for almost all } w \in \Omega. \tag{5.8}$$

From (5.8) we deduce that $\int |u - v|\mu_w(dv) \geq |u|$ for almost every $u \in \mathbf{C}$, and since

$$E\left(\int |u - v|\mu(dv)\mu(du)\right) = E\left(\int |u|\mu(du)\right),$$

we actually have

$$\int |u - v|\mu_w(dv) = |u|$$

for almost all $w \in \Omega$ and μ_w-almost all $u \in \mathbf{C}$. Let $w \in \Omega$ and $u \in \text{supp}\mu_w$; then

$$|u| = \int |u - v|\mu_w(dv).$$

Let z be a complex number of modulus 1 such that $z.u = |u|$. Then by (5.8) we have

$$|u| = \int |u - v|\mu_w(dv) \geq \int \text{Re}(zu - zv)\mu_w(dv)$$

$$= zu = |u|.$$

Hence $z(u - v) = |u - v|$ for μ_w-almost every $v \in \mathbf{C}$. Therefore the support of μ_w is localized in a real half-line issuing from u. Since this is true for any $u \in \text{supp}\mu_w$, it is clear that $\text{supp}\mu_w$ has at most two points, say $X(w)$ and $Y(w)$. Since μ_w is a probability on $\mathbf{C}$ we may write

$$\mu_w = (1 - \theta(w))\delta_{Y(w)} + \theta(w)\delta_{X(w)} \tag{5.9}$$

with $\theta(w) \in [0, 1]$. When $\theta(w) = 0$ or $\theta(w) = 1$, we have $X(w) = Y(w)$. Since $0 = \int u\mu_w(du)$, we deduce that $0 \in [X(w), Y(w)]$; in particular, on the set $\{X = Y\}$ we have $X = Y = 0$.

Let $x, y \in K$, so by (5.6) we have

$$E\left(\int \{\frac{1}{2}|x - u| + \frac{1}{2}|y - u| - |\frac{x+y}{2} - u|\}\mu(du)\right) = 0. \qquad (5.10)$$

Since $|\frac{x+y}{2} - u| \leq \frac{1}{2}|x - u| + \frac{1}{2}|y - u|$ is true for any $w \in \Omega$ and $u \in C$, by (5.10) we obtain

$$|\frac{x(w) + y(w)}{2} - u| = \frac{|x(w) - u|}{2} + \frac{|y(w) - u|}{2}$$

for almost every $w \in \Omega$ and μ_w-almost every $u \in C$. From (5.9) we deduce that $x(w)$ and $y(w)$ are both in half-lines issuing from $X(w)$ and $Y(w)$ respectively.

Since K is minimal for T, it is separable and therefore we can find a negligible set N (i.e. a set of measure 0) such that what we proved above would be true for any $w \notin N$ and any points x and y in K. In particular $x(w)$ belongs to a real line which passes through 0, because $0 \in K$, for any $x \in K$ and $w \notin N$. So one can deduce the conclusion of *(1)*.

To prove *(2)*, suppose that $K \subset L^1(\Omega, \mathcal{A}, P, \mathbf{R})$. We may assume that $X \leq 0 \leq Y$. On the set $\{X < Y\}$ we have $X < 0 < Y$, by the discussion following (5.9). Let $x \in K$; we have shown that 0 and $x(w)$ are in half-lines issuing from $X(w)$ and $Y(w)$, so $X(w) \leq x(w) \leq Y(w)$. When $X(w) = Y(w)$ all that can be deduced is that all $x(w)$ have the same sign, for any $x \in K$.

Let $\alpha \in \mathcal{D}$ and let X_α and Y_α be associated to (x_n^α). Fix $\alpha_0 \in \mathcal{D}$. On the set $\{X_{\alpha_0} < Y_{\alpha_0}\}$ we have $X_{\alpha_0} \leq x \leq Y_{\alpha_0}$ for all $x \in K$. Therefore

$$X_{\alpha_0} \leq x_n^\alpha \leq Y_{\alpha_0} \qquad (5.11)$$

holds for every $\alpha \in \mathcal{D}$ and every $n \in \mathbf{N}$. Using the properties of the support of μ^α and the random probability associated with (x_n^α), we obtain

$$X_{\alpha_0} \leq X_\alpha \leq Y_\alpha \leq Y_{\alpha_0}$$

for every $\alpha \in \mathcal{D}$. But if one considers the set $\{X_\alpha < Y_\alpha\} \cap \{X_{\alpha_0} < Y_{\alpha_0}\}$, a similar argument will give $X_\alpha = X_{\alpha_0}$ and $Y_\alpha = Y_{\alpha_0}$; and on the set $\{X_\alpha = Y_\alpha\} \cap \{X_\alpha < Y_\alpha\}$, the $x_n^{\alpha_0}(w)$ are on the same side of $X_\alpha(w)$, so $X_\alpha = Y_\alpha = X_{\alpha_0}$ or $X_\alpha = Y_\alpha = Y_{\alpha_0}$.

Finally, we see that for almost every $w \in \bigcup_{\alpha \in \mathcal{D}} \{X_\alpha < Y_\alpha\}$ the set $\{X_\alpha(w), Y_\alpha(w) : \alpha \in \mathcal{D}\}$ has at most two points. If $w \notin \bigcup_{\alpha \in \mathcal{D}} \{X_\alpha <$

Y_α}, we have $X_\alpha(w) = Y_\alpha(w)$ for all $\alpha \in \mathcal{D}$, and for any fixed $\alpha_0 \in \mathcal{D}$ and any $x \in K$, all $x(w)$ are in a half-line issuing from $X_{\alpha_0}(w)$. Hence the $X_\alpha(w)$ are in a half-line issuing from $X_{\alpha_0}(w)$ too, for any $\alpha \in \mathcal{D}$. Again, we can say that the set $\{X_\alpha(w) : \alpha \in \mathcal{D}\}$ has at most two points.

It follows that, for any $w \in \Omega$, there exist $U(w)$ and $V(w)$ such that

$$\text{supp}\mu_w \subset \{U(w), V(w)\} \tag{5.12}$$

for any random probability μ associated with any a.f.p.s. (x_n) in K. We may assume that $U(w) \leq V(w)$ for all $w \in \Omega$. Let $f(w, u) = 1 \wedge |u - U(w)| \wedge |v - V(w)|$; then

$$E\left(\int f(u)\mu(du)\right) = 0$$

because of (5.12). This will yield

$$\lim_{\mathcal{U}} E(|x_n - U| \wedge |x_n - V| \wedge 1) = 0.$$

Hence one can find a subsequence (x_{n_k}) such that

$$|x_{n_k} - U| \wedge |x_{n_k} - V| \wedge 1 \to 0$$

almost everywhere. Therefore $|x_{n_k} - U| \wedge |x_{n_k} - V|$ tends to 0 almost everywhere, so $|U| \wedge |V|$ is integrable, because $|U| \wedge |V| \leq \liminf_k |x_{n_k}|$.

Now let $f(w, u) = |u - U(w)| \wedge |v - V(w)|$; again we have $E(\int f(u)\mu(du)) = 0$, in particular $E(\int f(u)\mu^\alpha(du)) = 0$ for any $\alpha \in \mathcal{D}$. Thus *(2)* is proven. ∎

Theorem 5.5. *Any reflexive subspace R of $L^1(\Omega, \mathcal{A}, P)$ has the fixed point property.*

Proof. Assume the contrary. Then in R there is a nonempty weakly compact convex subset K which is minimal for a nonexpansive map T. By Proposition 5.4 *(1)*, we can reduce K to be formed by real functions. We may assume $\delta(K) = 1$. Let $\mathcal{U}$ be a nontrivial ultrafilter on $\mathbf{N}$, and as usual consider $\tilde{K}$ and $\tilde{T}$ in $(L^1)_{\mathcal{U}}$. We saw in the discussion following Definition 2.12 of Chapter 2 that $\tilde{K}$ can be seen as a subset of $L^1(\tilde{\Omega}, \tilde{\mathcal{A}}, \tilde{P})$, which was described in section I of

Chapter 2. Recall that if $\widetilde{A} \in \widetilde{\mathcal{A}}$ then there exists (A_n) in $\mathcal{A}$ such that $\widetilde{P}(\widetilde{A}) = \lim_{\mathcal{U}} P(A_n)$.

Let $\widetilde{x} = \widetilde{(x_n)}$ be a fixed point of $\widetilde{T}$ in $\widetilde{K}$. Proposition 5.4 *(2)* implies that

$$|\widetilde{x} - U| \wedge |\widetilde{x} - V| = 0$$

holds in $L^1(\widetilde{\Omega}, \widetilde{\mathcal{A}}, \widetilde{P})$. Therefore

$$\widetilde{x} = 1_{\underset{A}{\sim}} U + 1_{\underset{A^c}{\sim}} V \tag{5.13}$$

for some $\widetilde{A} \in \widetilde{\mathcal{A}}$. So for any $\widetilde{x}$ and $\widetilde{y}$ in $\text{Fix}(\widetilde{T})$, there exists $\widetilde{B}$ in $\widetilde{\mathcal{A}}$ such that

$$|\widetilde{x} - \widetilde{y}| = 1_{\underset{B}{\sim}} |U - V|. \tag{5.14}$$

Now let $\widetilde{x}$ and $\widetilde{y}$ be in $\text{Fix}(\widetilde{T})$ with $\|\widetilde{x} - \widetilde{y}\| = \delta(K) = 1$ (it is always possible to find such elements – see Theorem 3.1 *(2)*). Since $\text{Fix}(\widetilde{T})$ is metrically convex by Theorem 3.1 *(3)*, there exists $\widetilde{z}$ in $\text{Fix}(\widetilde{T})$ such that

$$\frac{1}{2}\|\widetilde{x} - \widetilde{y}\| = \|\widetilde{x} - \widetilde{z}\| = \|\widetilde{y} - \widetilde{z}\|.$$

Iterating this process, we can find, for every $n \in \mathbf{N}$, $\widetilde{x}_0, \ldots, \widetilde{x}_n \in \text{Fix}(\widetilde{T})$ such that

$$\|\widetilde{x}_n - \widetilde{x}_0\| = \sum_{i=1}^{n} \|\widetilde{x}_i - \widetilde{x}_{i-1}\|. \tag{5.15}$$

By (5.14), there exist $\widetilde{B}_i$, $i = 1, \ldots, n$, such that

$$|\widetilde{x}_i - \widetilde{x}_{i-1}| = 1_{\underset{B_i}{\sim}} |U - V| \text{ for } i = 1, \ldots, n.$$

Since $|\widetilde{x}_0 - \widetilde{x}_n| \leq \sum_{i=1}^{n} |\widetilde{x}_i - \widetilde{x}_{i-1}|$, equation (5.15) implies that $|\widetilde{x}_0 - \widetilde{x}_n| = \sum_{i=1}^{n} |\widetilde{x}_i - \widetilde{x}_{i-1}|$ $\widetilde{P}$-almost everywhere. Hence we may assume that $(\widetilde{B}_i)$ are mutually disjoint (otherwise one can take $\widetilde{B}_i \cap \{U \neq V\}$ instead of $\widetilde{B}_i$ for $i = 1, \ldots, n$).

Put $\widetilde{z}_i = \frac{\widetilde{x}_i - \widetilde{x}_{i-1}}{\|\widetilde{x}_i - \widetilde{x}_{i-1}\|}$ for $i = 1, \ldots, n$; then we clearly have

$$\|\sum_{i=1}^{n} \alpha_i \widetilde{z}_i\| = \sum_{i=1}^{n} |\alpha_i|$$

for any sequence of scalars (α_i). Therefore $\widetilde{R}$ isometrically contains ℓ_1^n for every $n \in \mathbb{N}$. This yields a contradiction, since R is super-reflexive [120]. ∎

Remarks 5.6. (1) In [46] one can find a version of Theorem 5.5 extended to certain Banach lattices. Indeed, let $(X, \|\cdot\|)$ be a Banach lattice. We say that $\|\cdot\|$ is *strictly monotone* (s.m.) if $\|u\| > \|v\|$ whenever $u \geq v \geq 0$ and $u \neq v$, or *uniformly monotone* (u.m.) if for all $\epsilon > 0$ there exists $\delta > 0$ such that $\|u\| \geq \|v\| + \delta$ whenever $u \geq v \geq 0$ and $\|u - v\| \geq \epsilon$, with $\|v\| = 1$.

It is easy to check that L_p $(1 \leq p < \infty)$ have u.m. norms. Also one can easily verify that $\widetilde{X}$ has s.m. norm if and only if X has u.m. norm. The generalization of Theorem 5.5 in [46] states

Theorem 5.7. *Let X be a Banach lattice with u.m. norms and assume that ℓ_1 is not finitely representable in X. Then X has the fixed point property.*

(2) In the proof of Theorem 5.5, we showed that for any $\widetilde{x}$ and $\widetilde{y}$ in $\mathrm{Fix}(\widetilde{T})$ and for any quasi-middle fixed point $\widetilde{z}$ of $\widetilde{x}$ and $\widetilde{y}$ (i.e. $\|\widetilde{x} - \widetilde{z}\| = \|\widetilde{y} - \widetilde{z}\| = \frac{1}{2}\|\widetilde{x} - \widetilde{y}\|$), we have

$$\|\widetilde{z} - \frac{\widetilde{x} + \widetilde{y}}{2}\| = \frac{1}{2}\|\widetilde{x} - \widetilde{y}\|.$$

Maurey [99] proved the following theorem, which can be seen as a generalization of this remark.

Theorem 5.8. *Suppose that K is a weakly compact convex set which is minimal for a nonexpansive map T. If there exists $\delta > 0$ such that for any two fixed points $\widetilde{x}$ and $\widetilde{y}$ of $\widetilde{T}$ in $\widetilde{K}$, there exists a quasi-middle fixed point $\widetilde{z}$ of $\widetilde{x}$ and $\widetilde{y}$ so that $\|\widetilde{z} - \frac{\widetilde{x}+\widetilde{y}}{2}\| \geq \frac{\delta}{2}\|\widetilde{x} - \widetilde{y}\|$. Then X is not super-reflexive.*

Since the proof of Theorem 5.8 is based on the notion of trees created by James [67], let us define this notion before proceeding.

Definition 5.9. A *tree* in X is a bounded family of vectors $\{x_{n,k} : n = 0, 1, \ldots \text{ and } k = 1, 2, 3, \ldots, 2^n\} \subset X$ satisfying $x_{n,k} = \frac{1}{2}(x_{n+1,2k-1} + x_{n+1,2k})$ for each $n \in \mathbb{N}$ and $k = 1, \ldots, 2^n$. A

δ-*tree* is a tree $(x_{n,k})$ satisfying $\|x_{n+1,2k-1} - x_{n+1,2k}\| \geq \delta$ for each $n \in \mathbb{N}$ and $k = 1, \ldots, 2^n$.

It is shown in [16] that a Banach space with the Radon-Nikodym property cannot contain a bounded δ-tree. Therefore any super-reflexive Banach space can't contain a bounded δ-tree.

Proof of Theorem 5.8. Without loss of generality, we may assume that there exist two fixed points $\tilde{x}_0$ and $\tilde{x}_1$ of $\tilde{T}$ with $\|\tilde{x}_0 - \tilde{x}_1\| = 1$. Let $\tilde{x}_{\frac{1}{2}}$ be a quasi-middle fixed point (in short, a q.m.f.p.) of $\tilde{x}_0$ and $\tilde{x}_1$ which satisfies

$$\|\tilde{x}_{\frac{1}{2}} - \frac{\tilde{x}_0 + \tilde{x}_1}{2}\| \geq \frac{\delta}{2}\|\tilde{x}_0 - \tilde{x}_1\| = \frac{\delta}{2}.$$

By the assumption, such a point does exist. Suppose for $n \in \mathbb{N}$ and $k = 0, 1, 2, \ldots, 2^n - 1$, $\tilde{x}_{\frac{k}{2^n}}$ is defined. Then choose a q.m.f.p. $\tilde{x}_{\frac{2k+1}{2^{n+1}}}$ of $\tilde{x}_{\frac{k}{2^n}}$ and $\tilde{x}_{\frac{k+1}{2^n}}$ such that

$$\|\tilde{x}_{\frac{2k+1}{2^{n+1}}} - \frac{1}{2}(\tilde{x}_{\frac{k}{2^n}} + \tilde{x}_{\frac{k+1}{2^n}})\| \geq \delta\|\tilde{x}_{\frac{k}{2^n}} - \tilde{x}_{\frac{k+1}{2^n}}\| = \frac{\delta}{2^{n+1}}.$$

Let $\tilde{z}_{n,k} = 2^n(\tilde{x}_{\frac{k}{2^n}} - \tilde{x}_{\frac{k-1}{2^n}})$ for $n \in \mathbb{N}$ and $k = 1, \ldots, 2^n$. Then $\|\tilde{z}_{n,k}\| = 1$ and

$$\begin{aligned}
\|\tilde{z}_{n+1,2k-1} - \tilde{z}_{n+1,2k}\| &= 2^{n+1}\|(\tilde{x}_{\frac{2k-1}{2^{n+1}}} - \tilde{x}_{\frac{2k-2}{2^{n+1}}}) - (\tilde{x}_{\frac{2k}{2^{n+1}}} - \tilde{x}_{\frac{2k-1}{2^{n+1}}})\| \\
&= 2^{n+1}\|2\tilde{x}_{\frac{2k-1}{2^{n+1}}} - (\tilde{x}_{\frac{2k-2}{2^{n+1}}} + \tilde{x}_{\frac{2k}{2^{n+1}}})\| \\
&= 2^{n+2}\|\tilde{x}_{\frac{2k-1}{2^{n+1}}} - \frac{1}{2}(\tilde{x}_{\frac{2k-2}{2^{n+1}}} + \tilde{x}_{\frac{2k}{2^{n+1}}})\| \\
&\geq 2^{n+2}\frac{\delta}{2^{n+2}} = \delta
\end{aligned}$$

for $n \in \mathbb{N}$ and $k = 1, \ldots, 2^n$. Hence $(\tilde{z}_{n,k})$ is a bounded δ-tree in $\tilde{X}$ Therefore X is not super-reflexive. ∎

Another application of Proposition 5.4 concerns the Hardy space H^1. H^1 consists of all holomorphic functions f on D, the open unit disc in $\mathbb{C}$, with norm

$$\|f\|_{H^1} = \lim_{r \to 1} \frac{1}{2\pi} \int |f(re^{it})|\, dt \quad \text{is finite.}$$

V. Maurey's Theorems

With this norm H^1 can be identified with a subspace of L^1. Note that if $f \in H^1$ and $f = \sum_{n=0}^{\infty} a_n e^{in\theta}$ then

$$\|f\|_{H^1} = \frac{1}{2\pi} \int_0^{2\pi} |\sum a_n e^{int}| \, dt.$$

More on the Hardy space can be found in [97].

Theorem 5.10. *The Hardy space H^1 has the fixed point property.*

Proof. Assume the contrary, and let K be a weakly compact convex subset of H^1 which is minimal for a nonexpansive map T. We may assume that $\delta(K) = 1$. Proposition 5.4 implies that there exist $x_0 \in K$ and a measurable function G, with $|G(t)| = 1$, such that $G(x - x_0)$ is a real function for any $x \in K$. By translation we can assume that $x_0 = 0$. Let $C = G \cdot K = \{G \cdot k : k \in K\}$; C is a weakly compact convex subset of $L_{\mathbf{R}}^1$ (the subspace of L^1 consisting of real functions). Let us show that C is compact under the given norm.

Let (x_n) be a sequence of elements of C which is weakly convergent to 0. We will prove that (x_n) converges to 0 in the norm. Consider the orthogonal projection $P : L^2 \to H^2$ defined, for any $f \in L^2$, by

$$P(f) = P(\sum_{-\infty}^{+\infty} a_k e^{ik\theta}) = \sum_{k=0}^{+\infty} a_k e^{ik\theta},$$

and the operators Q_n corresponding to Fejer's sums

$$Q_n(f) = \sum_{k=-2n}^{2n} (1 - \frac{|k|}{2n+1}) a_k e^{ik\theta}$$

for any $f \in L^2$.

It is known [97] that P is a continuous operator from L^1 into L^p for all $p < 1$. And for every $n \in \mathbf{N}$ and $q \in [1, \infty)$, Q_n is a continuous operator with norm less than 1, from L^q into L^q for $1 \le q < \infty$. Furthermore, $\|f - Q_n(f)\|_q \to 0$ as $n \to \infty$ for any $f \in L^q$.

Since (x_n) converges weakly to 0 in $L_{\mathbf{R}}^1$, it follows that $(\overline{G}x_n)$ converges weakly to 0 in H^1. Therefore the Fourier coefficients tend simply to 0. Hence we may assume that

$$\overline{G}x_n = e^{in\theta}b_n + c_n \tag{5.16}$$

98

with $b_n \in H^1$ for every $n \in \mathbb{N}$, and $\lim_{n \to \infty} \|c_n\|_1 = 0$, where $\|\cdot\|_1$ is the norm on L^1. (This can be seen as analogous to the conclusion of Proposition 0.14 of Chapter 0.) Notice that (b_n) is equi-integrable, because (x_n) is equi-integrable. Since x_n is a real function, we have by 5.16

$$x_n = e^{in\theta}Gb_n + Gc_n = e^{-in\theta}\overline{Gb_n} + \overline{Gc_n}$$

hence

$$b_n = e^{-2in\theta}\overline{G}^2\overline{b_n} + r_n \tag{5.17}$$

where $\lim_{n \to \infty} \|r_n\|_1 = 0$. Put $\overline{G}^2 = Q_{n-1}(\overline{G}^2) + L_n$; then $\|L_n\|_\infty \le 2$ for every $n \in \mathbb{N}$, and $\lim_{n \to \infty} \|L_n\|_q = 0$ for $q < \infty$. (This means that (L_n) converges to 0 in probability.) Since $P(e^{-2in\theta}Q_{n-1}(\overline{G}^2)\overline{b_n}) = 0$, we have by (5.17) that

$$b_n = P(b_n) = P(e^{-2in\theta}L_n\overline{b_n} + r_n).$$

On the other hand, we also have $\lim_{n \to \infty} \|L_n\overline{b_n}\|_1 = 0$, because $(\overline{b_n})$ is equi-integrable and (L_n) tends to 0 in probability, with $\|L_n\|_\infty \le 2$. Using the fact that P is continuous from L^1 to L^p for $p < 1$, we deduce that

$$\lim_{n \to \infty} \|b_n\|_p = 0.$$

And, because (b_n) is equi-integrable, we have

$$\lim_{n \to \infty} \|b_n\|_1 = 0,$$

and therefore by (5.16)

$$\lim_{n \to \infty} \|x_n\|_1 = 0.$$

This completes the proof of the statement made about C, which is therefore compact under the norm of L^1 defined above. It follows that K is compact under this norm; however, this produces a contradiction, since a minimal set which is compact must be a single point. This completes the proof. ∎

In the next theorem we give the proof of Maurey's result [61] concerning the existence of fixed points for isometries in super-reflexive Banach spaces. (It is still unknown whether super-reflexive Banach spaces have the fixed point property.)

Theorem 5.11. *Let C be a weakly compact convex subset of a super-reflexive Banach space X, and let $T : C \to C$ be an isometry, i.e. $\|Tx - Ty\| = \|x - y\|$ for all $x, y \in C$. Then T has a fixed point.*

Proof. By a result due to Enflo [63], we know that X is super-reflexive if and only if one can find an equivalent norm which is uniformly convex. Pisier [158] strengthened Enflo's result by proving that X has an equivalent norm $|\cdot|$ such that there exist $2 \le q < \infty$ and $0 < \gamma < \infty$ for which

$$| \frac{x+y}{2} |^q \le \frac{1}{2}(|x|^q + |y|^q) - \gamma^q |x - y|^q \tag{5.18}$$

for any $x, y \in X$. For simplicity we will assume that q=2 and later we will indicate the necessary modifications in the general case.

Assume to the contrary that $\text{Fix}(T) = \phi$. Then there exists a nonempty closed convex subset K of C which is minimal for T. We may assume as usual that $\delta(K) = 1$. Let us construct a function $\varphi : K \to [0, M]$ for some $M < \infty$ satisfying

$$\varphi(\frac{x+y}{2}) \ge \frac{1}{2}(\varphi(x) + \varphi(y)) + \frac{1}{2}\|x - y\|^2 \tag{5.19}$$

and

$$\varphi(T(x)) \ge \varphi(x) \tag{5.20}$$

for every $x, y \in K$.

Let $\tilde{K}$ and $\tilde{T}$ be defined in $\tilde{X} = (X)_u$ in the usual way. The equivalent norm $|\cdot|$ defined on X which satisfies (5.18) defines an equivalent norm on $\tilde{X}$ which satisfies the same inequality. Let $\tilde{f}$ be in $\text{Fix}(\tilde{T})$ and let $\mathcal{D}$ denote the set of dyadic rationals in $[0, 1]$, i.e.

$$\mathcal{D} = \{\frac{k}{2^n} : n \in \mathbb{N} \text{ and } k = 0, 1, \dots, 2^n\}.$$

For any $y \in K$, define a configuration C about y by

$$C = \{y_i^r : r \in \mathcal{D} \text{ and } i \in \{0, 1\}^{\mathbb{N}}\}$$

where $y_i^0 = y$, $y_i^1 = \tilde{f}$, and such that for $n \in \mathbb{N}$ and $k = 0, 1, \dots, 2^n$ we have

$$y_i^{\frac{k}{2^n}} = y_j^{\frac{k}{2^n}} \text{ if } i \uparrow n = j \uparrow n \tag{5.21}$$

and

$$y_i^{\frac{2k+1}{2^n}} \text{ is a metric midpoint of } y_i^{\frac{k}{2^{n-1}}} \text{ and } y_i^{\frac{k+1}{2^{n-1}}}. \tag{5.22}$$

This requires some explanation. First, if $i = (i_n)$ with $i_n \in \{0,1\}$, then $i \uparrow m = (i_1, i_2, \ldots, i_m)$. Condition (5.22) means that

$$\|y_i^{\frac{2k+1}{2^n}} - y_i^{\frac{k}{2^{n-1}}}\| = \|y_i^{\frac{2k+1}{2^n}} - y_i^{\frac{k+1}{2^{n-1}}}\| = \frac{1}{2}\|y_i^{\frac{k}{2^{n-1}}} - y_i^{\frac{k+1}{2^{n-1}}}\|.$$

Notice that $y_i^r \in \widetilde{K}$ and may not be in K – we will omit the "~" to keep the notation from getting too heavy!

By (5.21) we can see y_i^r with $i \in \{0,1\}^{\mathbb{N}}$, when $r = \frac{k}{2}$ and $k \le 2^n$. This means that the remainder does not count.

Figures (1) and (2) are given in order to make all these statements clear. Since $\|y - \tilde{f}\| = 1$, one can easily find the distance between the points in Figure (1) which are connected by lines. To a configuration $C = \{y_i^r\}$ we associate a sequence of positive numbers

$$\Delta(C) = \{\delta_i^r : i = (i_1, \ldots, i_{n-1}) \in \{0,1\}^{\mathbb{N}} \text{ and}$$
$$r = \frac{2k-1}{2^n} \text{ with } k = 1, \ldots, 2^{n-1} \text{ and } n \ge 1\} \tag{5.22}$$

where

$$\delta_i^r = \|y_{i,0}^r - y_{i,1}^r\|$$

and $i,0 = (i_1, \ldots, i_{n-1}, 0)$ and $i,1 = (i_1, \ldots, i_{n-1}, 1)$. Thus for example (see Figure (2))

$$\delta_0^{\frac{1}{2}} = \|y_0^{\frac{1}{2}} - y_1^{\frac{1}{2}}\| \text{ and } \delta_0^{\frac{3}{4}} = \|y_{00}^{\frac{3}{4}} - y_{01}^{\frac{3}{4}}\|.$$

We define the width of the configuration C by

$$W(C) = \sum_{\delta \in \Delta(C)} \delta^2. \tag{5.23}$$

Now we are ready to define the function φ. For each $y \in K$, put

$$\varphi(y) = \sup\{W(C) : C \text{ is a configuration about } y\}. \tag{5.24}$$

Let us prove that φ satisfies (5.19) and (5.20), and that there exists an M such that $\varphi(y) \leq M$ for all $y \in K$. First let us observe that for any $A, B, C, D \in \tilde{X}$, we have

$$4\gamma^2|D - B|^2 \leq |A - B|^2 + |B - C|^2 + |C - D|^2 + |D - A|^2 - |A - C|^2.$$
$$(5.24)$$

Indeed, put $E = \frac{C+A}{2}$, and since $2(E - D) = (C - D) - (D - A)$ and $\frac{C-A}{2} = C - E$, we have, by (5.18),

$$4\gamma^2|E - D|^2 \leq \frac{1}{2}\left(|C - D|^2 + |D - A|^2\right) - |C - E|^2$$

or

$$8\gamma^2|E - D|^2 \leq |C - D|^2 + |D - A|^2 - 2|C - E|^2.$$

Similarly we have

$$8\gamma^2|E - B|^2 \leq |A - B|^2 + |B - C|^2 - 2|A - E|^2.$$

Then since $|A - C|^2 = 2|C - E|^2 + 2|A - E|^2$, we obtain

$$8\gamma^2\left(|E - D|^2 + |E - B|^2\right) \leq |A - B|^2 + |B - C|^2 + |C - D|$$
$$+ |D - A|^2 - |A - C|^2;$$

but $|D - B|^2 \leq 2|E - D|^2 + 2|E - B|^2$, so we have derived (5.24).

Let $C = \{y_i^r\}$ be any configuration about $y \in K$ and let $\Delta(C) = \{\delta_i^r\}$. Define $\Delta'(C) = \{\beta_i^r\}$ where

$$\beta_i^r = |y_{i,0}^r - y_{i,1}^r|.$$

Since the two norms are equivalent, to prove that $W(C)$ is bounded by a number M it is enough to show that

$$W'(C) = \sum_{\beta \in \Delta'(C)} \beta^2$$

is bounded by a number independent of C. Note that $W'(C)$ can be seen as the width of the configuration C with respect to the norm $|\cdot|$.

Fix $m \geq 1$ and consider

$$4\delta^2 \sum_{\beta \in \Delta'_m(C)} \beta^2,$$

where $\Delta'_m(C) = \{\beta_{i_1,\ldots,i_{n-1}}^{\frac{2k-1}{2^n}}$ for $n \leq m\}$. Iterating (5.24) yields the inequality

$$4\delta^2 \sum_{\beta \in \Delta'_m(C)} \beta^2 \leq \lambda^2 - |\tilde{f} - y|^2,$$

where λ is such that $\frac{1}{\lambda}\|x\| \leq |x| \leq \lambda\|x\|$ holds for any $x \in X$. Since m was arbitrary, we deduce that

$$W(C) \leq \frac{\lambda^4}{4\gamma^2}.$$

So one can take $M = \frac{\lambda^4}{4\gamma^2}$.

We need to know that $\varphi(Tx) \geq \varphi(x)$ for any $x \in K$. This will follow from the fact that if $\{x_i^r\}$ is a configuration about x, then $\{\tilde{T}x_i^r\}$ is a configuration about Tx of the same width, since $\tilde{T}$ is an isometry. It remains to show that (5.19) holds for φ. Let x and y be in K and let $C_1 = \{x_i^r\}, C_2 = \{y_i^r\}$ be configurations about x and y. Also let $C = \{z_i^r\}$, where

$$z_{0,i}^{\frac{r+1}{2}} = \frac{1}{2}(x^1 + x_i^r)$$

$$z_{0,i}^{\frac{r}{2}} = \frac{1}{2}(x^0 + y_i^r)$$

$$z_{1,i}^{\frac{r+1}{2}} = \frac{1}{2}(y^1 + y_i^r)$$

$$z_{1,i}^{\frac{r}{2}} = \frac{1}{2}(y^0 + x_i^r).$$

Figure (3) should help to describe the configuration C obtained by combining C_1 and C_2.

It is clear that C is a configuration about $z = \frac{1}{2}(x+y)$. Therefore if we compute the width of C, the δ^2 terms of $W(C_1)$ and $W(C_2)$ are now divided by 4, but each appears twice. $W(C)$ also contains the extra term $\|z_0^{\frac{1}{2}} - z_1^{\frac{1}{2}}\|^2 = \|\frac{x-y}{2}\|^2$. Thus we have

$$W(C) = \frac{W(C_1) + W(C_2)}{2} + \|\frac{x-y}{2}\|^2$$

which implies inequality (5.19).

Now that φ has been constructed, we can complete the proof of Theorem 5.11 when $q = 2$. Let $0 < \epsilon < 1$ and set

$$K_\epsilon = \{x \in K : \varphi(x) \geq \sup \varphi - \epsilon\}.$$

V. Maurey's Theorems

Clearly K_ϵ is a nonempty subset of K which is invariant under T. Inequality (5.19) implies that $\overline{K_\epsilon}$ is convex, and also $\mathrm{diam}\overline{K_\epsilon} = \mathrm{diam}K_\epsilon \leq \epsilon\sqrt{2}$. The minimality of K implies that $\overline{K_\epsilon} = K$ for all $0 < \epsilon < 1$. This is a contradiction, since $\mathrm{diam}K = 1$.

Let us now indicate how this argument should be modified in the general case. The main difference is that (5.19) cannot be obtained if we define

$$W(C) = \sum_{\delta \in \Delta(C)} \delta^q.$$

Instead we must use weights, like so:

$$W_q(C) = \sum_{n=1}^{\infty} \sum_{k=1}^{2^{n-1}} \sum_{(i_1,\ldots,i_{n-1})\in\{0,1\}^{n-1}} 2^{n(q-2)}(\delta_{i_1,\ldots,i_{n-1}}^{\frac{2k-1}{2^n}}).$$

The argument is then the same. ∎

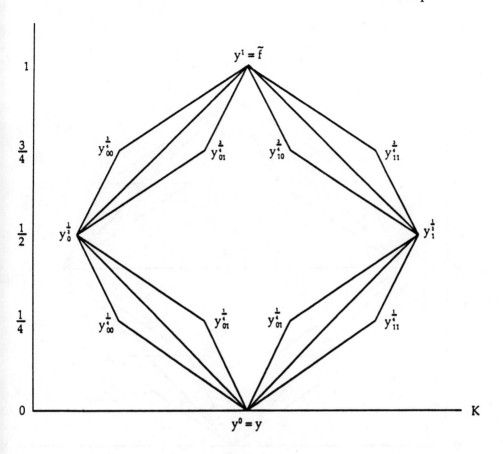

Figure 1

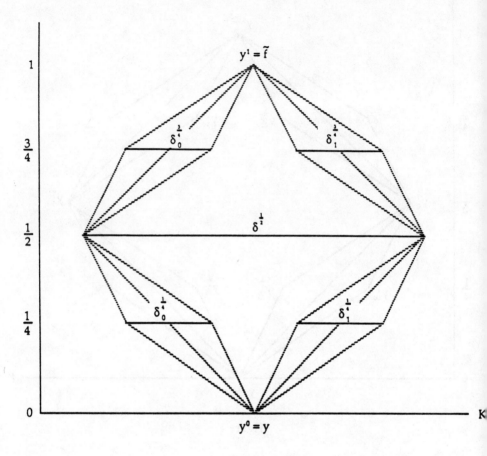

Figure 2

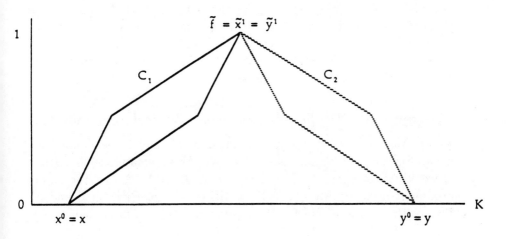

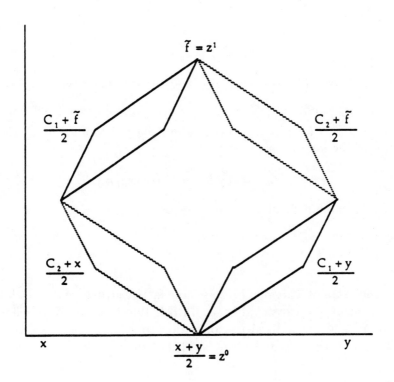

Figure 3

V. Maurey's Theorems

The next result concerns the Banach space c_0. The fixed point property was investigated in c_0 by the authors in [153] and also in [83], who obtained results for certain special domains; for example, they proved that the closed convex hull of a weakly convergent sequence has f.p.p. Whether or not c_0 has f.p.p. remained open until Maurey [149] gave a positive answer.

Theorem 5.12. *The Banach space c_0 has the fixed point property.*

Proof. Assume the contrary, and let $K, T, \widetilde{K}$, and $\widetilde{T}$ be defined as usual. Let (x_n) be an a.f.p.s. of T in K. We may suppose that (x_n) is weakly convergent to 0 and that $\operatorname{diam} K = 1$. Using the same argument as in the proofs of Proposition 0.14 of Chapter 0 and of Proposition 2.9, one can find a subsequence (x'_n) of (x_n) such that

$$\| |x'_n| \wedge |x_n| \| \le \frac{1}{n} \tag{5.25}$$

and

$$\lim_{n \to \infty} \|x_n - x'_n\| = 1. \tag{5.26}$$

One can deduce (5.26) from (5.25) and the fact that $\lim_{n \to \infty} \|x_n\| = 1$. Let $\widetilde{x} = \widetilde{(x_n)}$ and $\widetilde{y} = \widetilde{(x'_n)}$ be in $\widetilde{K}$. Then $\widetilde{x}$ and $\widetilde{y}$ are fixed points for $\widetilde{T}$. Let $\widetilde{z} = \widetilde{(z_n)}$ be a quasi-middle fixed point of $\widetilde{x}$ and $\widetilde{y}$; then by use of the lattice inequalities

$$|z_n| \le |z_n - x_n| \vee |z_n - x'_n| + |x_n| \wedge |x'_n| \tag{5.27}$$

one can obtain $\lim_{\mathcal{U}} \|z_n\| \le \frac{1}{2}$, since $\lim_{\mathcal{U}} \|x_n - z_n\| = \lim_{\mathcal{U}} \|x'_n - z_n\| = \frac{1}{2}$. This contradicts the facts that $\widetilde{z} \in \operatorname{Fix}(\widetilde{T})$ and $\|\widetilde{z} - 0\| = 1$, by Proposition 2.9. This completes the proof. ∎

The proof of Theorem 5.12 uses the lattice structure of c_0, which is induced by the canonical basis. In [24] Borwein and Sims refined the proof and generalized its conclusion to a large class of Banach lattices. Let us explain the central idea of their work.

Definition 5.13. Let X be a Banach lattice. Given a sequence (x_n) weakly convergent to x_0, we will say that (x_n) is *weakly orthogonal* if

$$\lim_{n \to \infty} \inf \lim_{m \to \infty} \inf \| |x_n - x_0| \wedge |x_m - x_0| \| = 0.$$

A subset C of X is a *weakly orthogonal* set if every weakly convergent sequence of points of C is weakly orthogonal. And we say X is *weakly orthogonal* if every weakly compact convex subset of X is weakly orthogonal.

In general it is not easy to prove that a given Banach lattice is weakly orthogonal. However, when the lattice has a good structure the verification is easier.

Definition 5.14. A Banach lattice X has the *Riesz approximation property* (R.A.P.) if there exists a family $\mathcal{P}$ of linear projections such that $P|x| = |P(x)|$ for all $P \in \mathcal{P}$,

$$P(X) \text{ is a finite dimensional ideal for all } P \in \mathcal{P}, \text{ and} \qquad (5.28)$$

$$\text{for each } x \in X, \ \inf_{P \in \mathcal{P}} \|x - P(x)\| = 0. \qquad (5.29)$$

($P(X)$ denotes the range of P.)

Proposition 5.15. *If X is a Banach lattice with R.A.P., then X is weakly orthogonal.*

Proof. We will prove the stronger result that for any $x \in X$ the mapping $y \mapsto |x| \wedge |y|$ is weak to norm continuous at 0. Let $\epsilon > 0$ and select $P \in \mathcal{P}$ so that $\||x| - P|x|\| \leq \epsilon$. Then if (y_α) is any net weakly convergent to 0, there is an α_0 such that for $\alpha \geq \alpha_0$,

$$\|P|y_\alpha|\| = \|P(y_\alpha)\| < \epsilon.$$

This is possible because the range of P is a finite dimensional subspace of X. Now, we have

$$|x| \wedge |y_\alpha| \leq P|x| \wedge P|y_\alpha| + \big| |x| - P|x| \big| \wedge |y_\alpha| + P|x| \wedge \big| |y_\alpha| - P|y_\alpha| \big|. \qquad (5.30)$$

Put $z = P|x| \wedge \big| |y_\alpha| - P|y_\alpha| \big|$; then $z \geq 0$ and z belongs to the ideal $P(X)$. Thus $z = P(z) = P|x| \wedge P \big| |y_\alpha| - P|y_\alpha| \big| = P|x| \wedge 0 = 0$, and so (5.30) reduces to

$$\begin{aligned}
|x| \wedge |y_\alpha| &\leq P|x| \wedge P|y_\alpha| + \big| |x| - P|x| \big| \wedge |y_\alpha| \\
&\leq P|x| \wedge P|y_\alpha| + \big| |x| - P|x| \big|,
\end{aligned} \qquad (5.31)$$

V. Maurey's Theorems

so that for $\alpha \geq \alpha_0$,

$$|||x| \wedge |y_\alpha||| \leq \||P|y_\alpha||| + |||x| - P|x||| \leq 2\epsilon.$$

The proof is complete. ∎

Taking $\mathcal{P}$ to be the standard basis projections, we have, for any set Γ, that the spaces $c_0(\Gamma)$ and $\ell_p(\Gamma)$ $(1 \leq p < \infty)$ have R.A.P. The spaces $\ell_\infty(\Gamma)$, $c(\Gamma)$, and $L_p[0,1]$ $(1 \leq p \leq \infty)$ fail to have the R.A.P. and, with the exception of $c(\Gamma)$, also fail to be weakly orthogonal. The Banach spaces X_β, discussed in 2.6, have the R.A.P. Before generalizing Theorem 5.12, let us give one definition.

Definition 5.16. The *Riesz angle* α of a Banach lattice X is defined by

$$\alpha(X) = \sup\{|||x| \vee |y||| : \|x\| \leq 1, \|y\| \leq 1\}. \qquad (5.32)$$

One can easily prove that $\alpha(X) = 1$ if and only if X is an M-space, and if X is an abstract L_p space $(1 \leq p \leq \infty)$, then $\alpha(X) = 2^{\frac{1}{p}}$. In particular $\alpha(c_0) = 1$. In the proof of Theorem 5.12, the inequality (5.27) was critical; a more general inequality involves the Riesz angle.

Proposition 5.17. *Let X be a Banach lattice with Riesz angle $\alpha(X)$. Then for any $x, y, z \in X$, we have*

$$\|z\| \leq \alpha(X)(\|x - z\| \vee \|y - z\|) + |||x| \wedge |y|||. \qquad (5.33)$$

Proof. We know

$$|z| \leq |x| + |x - z| \leq |x - z| \vee |y - z| + |x|,$$

and similarly

$$|z| \leq |x - z| \vee |y - z| + |y|.$$

Thus

$$|z| \leq |x - z| \vee |y - z| + |x| \wedge |y|,$$

and therefore

$$\|z\| \le \||x - z| \vee |y - z|\| + \||x| \wedge |y|\|.$$

Using the definition of $\alpha(X)$, we obtain the desired conclusion. ∎

Theorem 5.18. *Let X be a Banach lattice with Riesz angle $\alpha(X) < 2$ and let C be a weakly compact convex subset of X which is weakly orthogonal. Then C has the fixed point property.*

Proof. Assume the contrary, and let $T : C \to C$ be a nonexpansive mapping with empty fixed point set. Hence there exists a nonempty, closed, convex subset K of C which is minimal for T. We may assume $\delta(K) = 1$. Let (x_n) be an a.f.p.s. of T in K which is weakly convergent. As usual one can assume that $\text{weak} - \lim_{n \to \infty} x_n = 0$. Since C is weakly orthogonal, there exists a subsequence (x'_n) of (x_n) such that

$$\lim_{\substack{n_1 < n_2 \\ n_1 \to \infty}} \||x'_{n_1}| \wedge |x'_{n_2}|\| = 0.$$

Using Proposition 2.9, we may assume that

$$\lim_{\substack{n_1 < n_2 \\ n_1 \to \infty}} \|x'_{n_1} - x'_{n_2}\| = 1.$$

Let $\tilde{K}$, $\tilde{T}$ be in $(X)_{\mathcal{U}}$, where $\mathcal{U}$ is a nontrivial ultrafilter on $\mathbf{N}$. The inequality (5.33) can be written in $(X)_{\mathcal{U}}$ as

$$\|\tilde{z}\| \le \alpha(X)(\|\tilde{x} - \tilde{z}\| \vee \|\tilde{y} - \tilde{z}\|) + \||\tilde{x}| \wedge |\tilde{y}|\| \qquad (5.34)$$

for any $\tilde{x}, \tilde{y}, \tilde{z}$ in $(X)_{\mathcal{U}}$.

Put $\tilde{x} = \widetilde{(x'_n)}$ and $\tilde{y} = \widetilde{(x'_{n+1})}$; then

$$\|\tilde{x} - \tilde{y}\| = 1 \text{ and } \||\tilde{x}| \wedge |\tilde{y}|\| = 0. \qquad (5.35)$$

Let $\tilde{z}$ be a quasi-middle fixed point of $\tilde{x}$ and $\tilde{y}$. Then (5.34) gives us

$$\|\tilde{z}\| \le \alpha(X)(\frac{1}{2} \vee \frac{1}{2}) + 0 = \frac{\alpha(X)}{2},$$

and since $\|\tilde{z}\| = 1$, we deduce that $\alpha(X) \ge 2$. But this is a contradiction. ∎

It is not hard to see that $\alpha(X_\beta) = 1$ (see 2.6). And since X_β has R.A.P. for any $\beta > 0$, we deduce from Theorem 5.18 that X_β has f.p.p. Let us mention that this conclusion was also obtained by using Theorem 4.4. In both cases the proofs are more or less the same and use the same ideas.

VI. AN APPLICATION OF ULTRANETS

In this section we will give another example of a successful application of a non-standard method to fixed point problems. This material should be considered separate from the rest of this chapter, since the techniques used here are, in a sense, unrelated to the ultraproduct techniques discussed elsewhere. We should point out that ultranets have not found such power in fixed point theory as ultraproducts did with Maurey's and Lin's results.

The use of net was motivated when an approach using sequences failed specifically because the topological spaces involved were not of a "sequential" nature (see [33, 34, 133]). Recently Kirk [115] used ultranets to study the fixed point problem in product spaces (see [115, 117, 120, 121, 122, 123, 138, 196]). We will describe Kirk's approach in the hope that the method may motivate new results in fixed point theory.

In Chapter 1 there is a review of the theory of nets and ultranets. Let us begin by showing how one can naturally associate a net of mappings to a given map. Let K be a weakly compact convex subset of a Banach space X and consider a nonexpansive mapping $T : K \to K$. Put $T^0 = I$ and $T^{n+1} = TT^n$ for $n \in \mathbb{N}$. This is the classical sequence of iterates of T. Now let us extend this iteration to the ordinals. We proceed by transfinite induction. Let β be any ordinal and assume that T^α has been defined for all $\alpha < \beta$. There are two possibilities, that $\beta = \bar{\beta} + 1$ (β is a successor ordinal) or that no such $\bar{\beta}$ exists (β is a limit ordinal). In the first case, simply set $T^\beta = TT^{\bar{\beta}}$. In the second case, we will need to use ultranets. By Theorme 3.7 of Chapter 1, the net $\{\alpha : \alpha < \beta\}$ has a universal subnet (that is, a subnet which is an ultranet) which we will denote by $\{\alpha_i : i \in I_\beta\}$. Since K is weakly compact, by the same theorem we have

$$T^\beta(x) = \text{weak} - \lim_{i \in I_\beta} T^{\alpha_i}(x) \qquad (6.1)$$

exists for any $x \in K$. One can check that since the norm is weak-lower-semi-continuous, T^β is nonexpansive. Therefore, by transfinite induction, we can construct a nonexpansive mapping T^β for any ordinal β.

Let $x \in K$ and consider the transfinite orbit $(T^\alpha(x))_{\alpha \in \mathcal{O}}$, where $\mathcal{O}$ denotes the class of all ordinals. One would expect that this orbit will either approach $\text{Fix}(T)$ or eventually become constant. In fact this will occur if X has a "good" norm, such as the Kadec-Klee norm [86].

Definition 6.1. A Banach space X has the *Kadec-Klee norm* (in short the K.K. norm) if for any (x_n) in X with $\|x_n\| \leq 1$ and $\text{sep}\{x_n\} = \inf\{\|x_n - x_m\| : n \neq m\} > 0$, we have $\|x\| < 1$, where $x = \text{weak} - \lim_{n \to \infty} x_n$.

It is not hard to check that uniformly convex norms are K.K. norms. For more on K.K. norms, one can consult [86, 121]. Interestingly, even if this notion is a weak-geometric condition, it yields the following result.

Proposition 6.2. *Suppose X is a Banach space which has the K.K. norm and let K be a weakly compact convex subset of X. Also suppose $\mathcal{D}$ is a directed set and that $\{x_\alpha : \alpha \in \mathcal{D}\}$ is a net in K which for some $p \in X$ satisfies $\|x_\alpha - p\| = r > 0$ for all $\alpha \in \mathcal{D}$. Then if $\{x_\alpha\}$ converges weakly but not strongly to $x \in K$, we have*

$$\|x - p\| < r.$$

Proof. We may assume $r = 1$ and $p = 0$, and we suppose that $\|x\| = 1$. Let

$$K_\alpha = \overline{\text{conv}}\{x_\beta : \beta \geq \alpha\}$$

for each $\alpha \in \mathcal{D}$. K_α is a weakly compact convex subset of K and contains x, so there exists $y_\alpha \in K_\alpha$ such that

$$\|y_\alpha\| = \inf\{\|z\| : z \in K_\alpha\}.$$

Obviously we have $\text{weak} - \lim_{\alpha \in \mathcal{D}} y_\alpha = x$. Moreover, $\{\|y_\alpha\| : \alpha \in \mathcal{D}\}$ is an increasing net of real numbers, with

$$\lim_{\alpha \in \mathcal{D}} \|y_\alpha\| = 1, \tag{6.2}$$

113

since the norm is weak-lower-semi-continuous. On the other hand, since $\{x_\beta : \beta \geq \alpha\}$ is not pre-compact for this norm, it contains a sequence which is weakly convergent but not convergent. Because the norm is K.K., this implies that

$$\|y_\alpha\| = \inf\{\|z\| : z \in K_\alpha\} < 1.$$

By (6.2), one can find $\alpha_n \in \mathcal{D}$ for any $n \in \mathbf{N}$ such that

$$\|y_{\alpha_n}\| \geq 1 - \frac{1}{n}. \tag{6.3}$$

Assume first that there exists an $\alpha \in \mathcal{D}$ which is greater than or equal to all the α_n's. By (6.3), this implies that $\|y_\alpha\| \geq 1 - \frac{1}{n}$, which is a contradiction since $\|y_\alpha\| < 1$. Hence for any $\alpha \in \mathcal{D}$ there is some α_n greater than it. But this implies that (y_{α_n}) is weakly convergent (to x), and our assumption on the norm will imply $\|x\| < 1$. This contradiction completes the proof. ∎

This proposition will enable us to prove a result whose sequential analogue fails even in Hilbert spaces [70]. We denote the set of countable ordinals by Ω_0.

Theorem 6.3. *Let X be a Banach space with K.K. norm, K a weakly compact convex subset of X, and $T : K \to K$ a nonexpansive map with nonempty fixed point set. Let $F = \frac{I+T}{2}$ and consider the transfinite iterations (F^α) of F. Then for each $x \in K$, there exists $\alpha \in \Omega_0$ such that*

$$F^\alpha(x) \in \mathrm{Fix}(T).$$

Proof. Let $p \in \mathrm{Fix}(T)$ and $x \in K$. Put $r_\alpha = \|F^\alpha(x) - p\|$. Since p is a fixed point, $\{r_\alpha : \alpha \in \Omega_0\}$ is a decreasing net of positive numbers. Let $r = \lim_{\alpha \in \Omega_0} r_\alpha = \inf_{\alpha \in \Omega_0} r_\alpha$. Then for any $n \in \mathbf{N}$, there exists $\alpha_n \in \Omega_0$ such that $\|F^{\alpha_n}(x) - p\| \leq r + \frac{1}{n}$. Since $\alpha = \sup_n \alpha_n \in \Omega_0$, we have $\|F^\alpha(x) - p\| = r$. Therefore we may assume $r > 0$ since otherwise the conclusion follows. Clearly, for any $\beta \geq \alpha$, $\beta \in \Omega_0$, we have $\|F^\beta(x) - p\| = r$; and in particular, if β is a limit ordinal, then by (6.1)

$$F^\beta(x) = \mathrm{weak} - \lim_{i \in I_\beta} F^{\alpha_i}(x).$$

Using the fact that $\{\alpha_i : i \in I_\beta\}$ is a universal subnet of $\{\gamma \in \Omega_0 : \gamma < \beta\}$, one can find $i_0 \in I_\gamma$ such that $\alpha_i \geq \alpha$ for $i \geq i_0$. So

$$F^\beta(x) = \text{weak} - \lim_{\substack{i \in I_\beta \\ i \geq i_0}} F^{\alpha_i}(x).$$

Since $\|F^{\alpha_i}(x) - p\| = r$ for $i \geq i_0$, Proposition 6.2 implies that $F^\beta(x)$ is the strong limit of $\{F^{\alpha_i}(x) : i \in I_\beta, i \geq i_0\}$. Therefore, because

$$F^{\beta+1}(x) = FF^\beta(x) = F(\lim_{\substack{i \in I_\beta \\ i \geq i_0}} F^{\alpha_i}(x)) = \lim_{\substack{i \in I_\beta \\ i \geq i_0}} F^{\alpha_i+1}(x)$$

and

$$\lim_{\substack{i \in I_\beta \\ i \geq i_0}} \|F^{\alpha_i+1}(x) - F^{\alpha_i}(x)\| = 0,$$

we get

$$F^\beta(x) \in \text{Fix}(F) = \text{Fix}(T).$$

And we are done. ∎

The next result concerns a problem which is different in nature but which uses the same techniques. Let us first recall the product fixed point problem. Suppose $(X, \|\cdot\|_X)$ and $(Y, \|\cdot\|_Y)$ are Banach spaces and let $X \oplus Y$ denote the product space of X and Y. Consider any norm $\|\cdot\|$ on $X \oplus Y$. It is unknown whether $(X \oplus Y, \|\cdot\|)$ has the fixed point property whenever $(X, \|\cdot\|_X)$ and $(Y, \|\cdot\|_Y)$ have f.p.p.

(Concerning a generalization of this problem to metric spaces and to certain categories of mappings, the answer is well known [64].)

As an approach to the original problem, one can consider weakly compact convex subsets of $X \oplus Y$ obtained as the product of weakly compact convex sets K_1, K_2 which are subsets of X and Y respectively. Even in this case, the answer is unknown in general, despite some positive results [115, 117, 120]. Since the answer is already known [106, 122] when the norm product is

$$\|(x,y)\|_p = (\|x\|_X^p + \|y\|_Y^p)^{\frac{1}{p}}$$

for $1 \leq p < \infty$, we will apply the ultranet technique to the case $p = \infty$, i.e. when

$$\|(x,y)\|_\infty = \max(\|x\|_X, \|y\|_Y). \tag{6.4}$$

VI. An Application of Ultranets

Theorem 6.4. *Let X and Y be two Banach spaces and suppose that X has the K.K. norm. Let K_1 be a weakly compact convex subset of X and K_2 a subset of Y. Assume that K_1 and K_2 have the fixed point property for nonexpansive mappings. Then any mapping $T : K_1 \oplus K_2 \to K_1 \oplus K_2$ which is nonexpansive (with respect to $\|\cdot\|_\infty$) has a fixed point.*

<u>*Proof.*</u> Let P_1 and P_2 denote the natural projections of $X \oplus Y$ onto X and Y, respectively. Fix $y \in K_2$ and define $T_y : K_1 \to K_1$ by

$$T_y = P_1 \circ T.$$

It is clear that T_y is nonexpansive. Let $F_y = \frac{I+T_y}{2}$; then from Theorem 6.3 we have that for any fixed $x_0 \in K_1$, there exists $\alpha \in \Omega_0$ such that $F_y^\alpha(x_0) \in \mathrm{Fix}(T_y)$. Put $p_y = F_y^\alpha(x_0)$. Let us show that the map $y \mapsto p_y$ is nonexpansive. For $u, v \in K_2$ we have

$$\|F_u(x_0) - F_v(x_0)\|_X = \frac{1}{2}\|T_u(x_0) - T_v(x_0)\|_X$$

$$\leq \frac{1}{2}\|T(x_0, u) - T(x_0, v)\|_X$$

$$\leq \|u - v\|_Y.$$

We make the inductive assumption that

$$\|F_u^\beta(x_0) - F_v^\beta(x_0)\|_X \leq \|u - v\|_Y \tag{6.5}$$

for all $\beta \in \Omega_0$, $\beta < \alpha$.

If α is an ordinal limit, the weak-lower-semi-continuity of the norm will imply inequality (6.5) for $\beta = \alpha$. Therefore we assume that $\alpha = \bar{\alpha} + 1$. Hence

$$\|F_u^\alpha(x_0) - F_v^\alpha(x_0)\|_X = \|F_u F_u^{\bar\alpha}(x_0) - F_v F_v^{\bar\alpha}(x_0)\|_X$$

$$\leq \frac{1}{2}\|T_u F_u^{\bar\alpha}(x_0) - T_v F_v^{\bar\alpha}(x_0)\|_X +$$

$$\frac{1}{2}\|F_u^{\bar\alpha}(x_0) - F_v^{\bar\alpha}(x_0)\|_X$$

$$= \frac{1}{2}\|P_1 T(F_u^{\bar\alpha}(x_0), u) - P_1 T(F_v^{\bar\alpha}(x_0), v)\|_X +$$

$$\frac{1}{2}\|F_u^{\bar\alpha}(x_0) - F_v^{\bar\alpha}(x_0)\|_X$$

$$\leq \frac{1}{2}\|u - v\|_Y + \frac{1}{2}\|u - v\|_Y,$$

since T is $\|\cdot\|_\infty$-nonexpansive and

$$\|F_u^{\bar\alpha}(x_0) - F_v^{\bar\alpha}(x_0)\|_X \le \|u - v\|_Y.$$

Therefore inequality (6.5) holds for successor ordinals. Hence we have proven that (6.5) is true for all $\beta \in \Omega_0$, and we can then deduce that the mapping $y \mapsto p_y$ is nonexpansive, i.e.

$$\|p_y - p_z\|_X \le \|y - z\|_Y \text{ for any } y, z \in K_2. \tag{6.6}$$

Now define $G : K_2 \to K_2$ by

$$G(y) = P_2 T(P_y, y) \text{ for any } y \in K_2.$$

It is clear from (6.6) that G is nonexpansive and therefore has a nonempty fixed point set. Let $y \in \text{Fix}(G)$; then clearly

$$T(p_y, y) = (p_y, y).$$

We have found a point of $K_1 \oplus K_2$ which is left fixed by T. ∎

Bibliography

[1] A. Aksoy, M. Nakamura, The approximation numbers $\gamma_n(T)$ and Q-precompactness, *Mathematica Japonica* (6) **31** (1986), 827-840.

[2] D. E. Alspach, A fixed point free nonexpansive map, *Proc. Amer. Math. Soc.* **82** (1981), 423-424.

[3] D. Amir, On Jung's constant and related constants in normed linear spaces, in *Longhorn Notes, Texas Functional Analysis Seminar*, The University of Texas, 1982-1983, 143-160.

[4] A. Andrew, Spreading basic sequences and subspaces of James' quasi reflexive space, *Math. Scand.* **48** (1981), 108-118.

[5] A. Andrew, James' quasi reflexive space is not isomorphic to any subspace of its dual, *Israel J. Math.* **38** (1981), 276-282.

[6] N. Aronszajn, P. Panitchpakdi, Extension of uniformly continuous transformations and hyperconvex metric spaces, *Pac. J. Math.* **6** (1956), 405-439.

[7] K. Astala, On measures of noncompactness and ideal variations in Banach spaces, *Ann. Acad. Sci. Fenn. Ser. A.I. Math. Diss.* **29** (1980), 1-42.

[8] J. S. Bae, Reflexivity of a Banach space with a uniformly normal

structure, *Proc. Amer. Math. Soc.* **90** (1984), 269-270.

[9] J. B. Baillon, Quelques aspects de la theorie des points fixes dans les espaces de Banach I, in *Seminaire d'Analyse Fonctionelle*, Ecole Polytechnique, Paris, 1978-1979.

[10] J. B. Baillon, Nonexpansive mapping and hyperconvex spaces, *Amer. Math. Soc. Contemp. Math.* vol.**72** (1986), 11-20.

[11] J. B. Baillon, R. Schönberg, Asymptotic normal structure and fixed points of nonexpansive mappings, *Proc. Amer. Math. Soc.* **81** (1981), 257-264.

[12] S. Banach, Sur les operations dans les ensembles abstraits et leures applications, *Fund. Math.* **3** (1922), 133-181.

[13] S. Banach, S. Saks, Sur la convergence forte dans les espaces L^p, *Studia Math.* **2** (1930), 51-57.

[14] J. Banas, On modulus of noncompact convexity and its properties, *Canad. Math. Bull.* (2) 30 (1987), 186-192.

[15] J. Banas, K. Goebel, Measure of noncompactness in Banach spaces, in *Lecture Notes in Pure and Applied Mathematics*, Vol. 60, ed. by Marcel Dekker, New York, 1980.

[16] B. Beauzamy, *Introduction to Banach Spaces and Their Geometry*, Math. Studies **68**, North-Holland, 1982.

[17] B. Beauzamy, Banach-Saks properties and spreading models, *Math. Scand.* **44** (1979), 357-384.

[18] B. Beauzamy, J. T. Lapreste, *Modeles Etales des Espaces de Banach*, Hermann, Paris.

[19] A. Beck, A convexity condition in Banach spaces and the strong law of large numbers, *Proc. Amer. Math. Soc.* **13** (1962), 329-334.

[20] S. Bellenot, Transfinite duals of quasi reflexive Banach spaces, *Trans. Amer. Math. Soc. 273* (1981), 276-282.

[21] L. P. Belluce, W. A. Kirk, Nonexpansive mappings and fixed points in Banach spaces, *Illinois J. Math.* **11** (1967), 474-479.

[22] L. P. Belluce, W. A. Kirk, E. F. Steiner, Normal structure in Banach spaces, *Pacific J. Math.* **26** (1968), 433-440.

[23] J. Bernal, F. Sullivan, Multidimensional volumes, superreflexivity and normal structure in Banach spaces, *Illinois J. Math.* **27** (1983), 501-513.

[24] J. M. Borwein, B. Sims, Nonexpansive mappings on Banach lattices and related topics, *Houston J. Math.* **10** (1984), 339-356.

[25] N. Bourbaki, *Element de Mathematiques: Topologie Generale*, Hermann, Paris, 1953.

[26] J. Bretagnolle, D. Dacunha-Castelle, J. L. Krivine, Lois stables et espaces L^p, *Ann. Inst. Henri Poincare, Sect. B* 2 (1966), 231-259.

[27] M. S. Brodskii, D. P. Milman, On the center of a convex set. *Dokl. Akad. Nauk. SSSR 59* (1948), 837-840 (Russian).

[28] L. E. J. Brouwer, Uber abbildungen von mannigfaltigkeiten, *Math. Ann.* **71** (1912), 97-115.

[29] L. E. J. Brouwer, An intuitionist correction of the fixed point theorem on the sphere, *Proc. Royal Soc. London (A)* **213** (1952), 1-2.

[30] F. E. Browder, Fixed point theorems for noncompact mappings in Hilbert space, *Proc. Nat. Acad. Sci. U.S.A.* **53** (1965), 1272-1276.

[31] F. E. Browder, Nonexpansive nonlinear operators in a Banach

space, *Proc. Nat. Acad. Sci. U.S.A.* **54** (1965), 1041-1044.

[32] R. E. Bruck, A simple proof of the mean ergodic theorem for Banach spaces, *Israel J. Math.* **32** (1979), 107-116.

[33] R. E. Bruck, A common fixed point theorem for a commuting family of nonexpansive mappings, *Pacific J. Math.* **53** (1974), 59-71.

[34] R. E. Bruck, A common fixed point theorem for compact convex semigroups of nonexpansive mappings, *Proc. Amer. Math. Soc.* (1975), 113-116.

[35] A. Brunel, L. Sucheston, On B-convex Banach spaces, *Math. Systems Theory* **7** (1974), 294-299.

[36] A. Brunel, L. Sucheston, On J-convexity and ergodic super-properties of Banach spaces, *Trans. Amer. Math. Soc.* **204** (1975), 79-90.

[37] W. L. Bynum, Normal structure in Banach spaces, *Manuscripta Math.* **11** (1974), 203-209.

[38] W. L. Bynum, A class of spaces lacking normal structure, *Comp. Math.* **25** (1972), 233-236.

[39] W. L. Bynum, Normal structure coefficients for Banach spaces, *Pac. J. Math.* **86** (1980), 427-436.

[40] P. G. Casazza, *Tzirelson's spaces*, Dept. of Math, Univ. of Alabama, 1982.

[41] P. G. Casazza, James' quasi-reflexive is primary, *Israel J. Math.* **26** (1977), 294-305.

[42] P. G. Casazza, W. B. Johnson, L. T. Tzafriri, Tzirelson's space, Preprint.

Bibliography

[43] P. G. Casazza, B. L. Lin, R. H. Lohman, On nonreflexive Banach spaces which contain no c_0 or ℓ_p, *Can. J. Math.* **32-6** (1980), 1382-1389.

[44] P. G. Casazza, R. H. Lohman, A general construction of spaces of the types of R. C. James, *Can. J. Math.* **27** (1975), 1263-1270.

[45] P. G. Casazza, E. Odell, Tzirelson's space II, Preprint.

[46] P. G. Casazza, T. J. Shura, Tzirelson's space, Preprint.

[47] A. L. Cauchy, *Leçon sur les calculs differentiel et integral, Vol. 1 and 2*, Paris, 1884.

[48] C. C. Chang, H. J. Keisler, *Model Theory*, North-Holland, 1973.

[49] S. Chutao, M. A. Khamsi, W. M. Kozlowski, Some geometrical properties and fixed point theorems in Orlicz modular spaces, Preprint.

[50] J. A. Clarkson, Uniformly convex spaces, *Trans. Amer. Math. Soc.* **40** (1936), 396-414.

[51] D. Dacunha-Castelle, Ultraproduits d'espaces de Banach, in *Seminaire Goulaouic-Schwartz*, 1971-1972, Exposes IX, X.

[52] D. Dacunha-Castelle, J. L. Krivine, Applications des ultraproduits a l'etude des espaces et des algebres de Banach, *Studia Math.* **41** (1972), 315-334.

[53] M. M. Day, *Normed Linear Spaces*, Springer-Verlag.

[54] M. M. Day, R. C. James, S. Swaminathan, Normed linear spaces that are uniformly convex in every direction, *Canad. J. Math.* **23** (1971), 1051-1059.

[55] J. Diestel, J. J. Uhl, *Vector Measures*, American Mathematical Society, Providence, 1977.

[56] D. van Dulst, Equivalent norms and the fixed-point property for nonexpansive mappings, *J. London Math. Soc.* **25** (1982), 139-144.

[57] D. van Dulst, B. Sims, Fixed points of nonexpansive mappings and Chebyshev centers in Banach spaces with norms of type (KK), Preprint.

[58] D. van Dulst, D. Voulk, (KK)-properties, normal structure and fixed points of nonexpansive mappings in Orlicz sequence spaces, *Canad. J. Math.* (3) **38** (1986), 728-750.

[59] A. Dvoretzky, Some results on convex bodies and Banach spaces, in *Proc. Symp. on Linear Spaces*, Jerusalem, 1961, 123-160.

[60] I. Ekeland, Sur les problemes variationnelles, *C. R. Paris, Ser. A. B.* **275** (1972), 1057-1059.

[61] J. Elton, P. K. Lin, E. Odell, S. Szareck, Remarks on the fixed point problem for nonexpansive maps, *Contemp. Math. Amer. Math. Soc.* **18** (1983), 87-120.

[62] P. Enflo, A counterexample to the approximation problem in Banach spaces, *Acta. Math.* **130** (1973), 309-317.

[63] P. Enflo, Banach spaces which can be given an equivalent uniformly convex norm, *Israel J. Math.* **13** (1972), 281-288.

[64] E. R. Fadell, Recent results in the fixed-point theory of continuous maps, *Bull. Amer. Math. Soc.* **76** (1970), 10-29.

[65] H. Fakhouri, Directions d'uniforme convexité dans un espace normé, in *Seminaire Choquet (Initiation á l'analyse) 14 anneé* (6), 1974-1975, 1-16.

[66] T. Fiegel, On a recent result of G. Pisier, in *Texas Functional Analysis Seminar*, The University of Texas, Austin, 1982-1983, 1-14.

[67] T. Fiegel, W. B. Johnson, A uniformly convex space which contains no ℓ_p, *Compositio Math.* **29** (1974), 179-190.

[68] M. Furi, M. Martelli, On the minimal displacement of points under α-lipschitz maps in normed spaces, *Boll. Un. Mat. Ital.* **4-9** (1974), 791-799.

[69] A. L. Garkavi, The best possible net and the best possible cross-section of a set in a normed space, *Amer. Math. Soc. Trans., Series 2* **39** (1964), 111-132.

[70] A. Genel, J. Lindenstrauss, An example concerning fixed points, *Israel J. of Math.* **22-1** (1975), 81-86.

[71] J. R. Giles, B. Sims, S. Swaminathan, A geometrically aberrant Banach space with normal structure, *Bull. Aust. Math. Soc.* **31** (1985), 75-81.

[72] A. A. Gillespie, B. B. Williams, Fixed-point theorems for nonexpansive mappings in Banach spaces with uniformly normal structure, *Applicable Anal.* **9** (1979), 121-124.

[73] K. Goebel, On the structure of minimal invariant sets for nonexpansive mappings, *Ann. Univ. Mariae Curie-Sklodowska* **29** (1975), 73-77.

[74] K. Goebel, On the minimal displacement of points under lipschitzian mappings, *Pacific J. Math.* **45** (1973), 151-163.

[75] K. Goebel, W. A. Kirk, *Topics in Metric Fixed Point Theory* (to appear in Cambridge University Press).

[76] K. Goebel, T. Komorowski, Retracting balls onto spheres and minimal displacement problems, Preprint.

[77] K. Goebel, S. Reich, *Uniform convexity, hyperbolic geometry and nonexpansive mappings*, Dekker, New York and Basel, 1984.

[78] K. Geobel, T. Sekowski, The modulus of noncompact convexity, *Ann. Univ. Mariae Curie-Sklokowska* **38** (1984), 41-48.

[79] D. Göhde, Zum Prinzip der kontraktiven abbildung, *Math. Nachr.* **30** (1965), 251-258.

[80] J. P. Gossez, E. Lami-Dozo, Normal structure and Schauder bases, *Acad. Roy. Belgique, Bulletin des Sciences* **55** (1969), 673-681.

[81] J. P. Gossez, E. Lami-Dozo, Some geometrical properties related to the fixed point theory for nonexpansive mappings, *Pac. J. Math.* **40** (1972), 565-573.

[82] S. Guerre, J. T. Lapreste, Quelques proprietes des modeles etales sur les espaces de Banach, *Ann. Inst. Henri Poincare, Nouv. Ser. Sect. B* **16-4** (1980), 339-347.

[83] R. Haydon, E. Odell, Y. Steinfeld, A fixed point theorem for a class of star shaped sets in c_0.

[84] S. Heinrich, Ultraproduct in Banach space theory, *J. Reine and Ang. Mat.* **313** (1980), 72-104.

[85] C. W. Henson, Nonstandard hulls of Banach spaces, *Israel J. Math* **25** (1976), 108-144.

[86] R. E. Huff, Banach spaces which are nearly uniformly convex, *Rocky Mountain J. of Math.* **10** (1980), 743-749.

[87] R. Isbell, Six theorems about injective metric spaces, *Comment. Math. Helv.* **39** (1964), 65-76.

[88] V. I. Istratescu, Fixed point theory, *Mathematics and its Application* **7**, Reidel, 1979.

[89] R. C. James, Some self-dual properties of normed linear spaces, *Ann. Math. Studies* **69** (1972), 159-175.

[90] R. C. James, Uniformly non-square Banach spaces, *Ann. Math.* **80** (1964), 542-550.

[91] R. C. James, Bases and reflexivity of Banach spaces, *Ann. of Math.* **52** (1950), 518-527.

[92] R. C. James, A non-reflexive Banach space isometric to its second conjugate, *Proc. Nat. Sci. U.S.A.* **37** (1951), 174-177.

[93] R. C. James, Banach spaces quasi-reflexive of order one, *Studia Math.* **60** (1977), 157-177.

[94] R. C. James, Super-reflexive spaces with bases, *Pac. J. Math.* **41-2** (1972), 409-419.

[95] E. Jawhari, D. Misane, M. Pouzet, Retracts, graphs, and ordered sets from the metric point of view, *Publications du Dept. de Math., Universite Claude Bernard*, Lyons.

[96] L. A. Karlovitz, Existence of a fixed point for a nonexpansive map in a space without normal structure, *Pacific J. Math* **66** (1976), 153-159.

[97] Y. Katznelson, *An Introduction to Harmonic Analysis*, John Wiley & Sons, New York, 1968.

[98] J. L. Kelley, *General Topology*, Van Nostrand, 1955.

[99] M. A. Khamsi, James quasi reflexive space has the fixed point property, *Bull. Austral. Math. Soc.* **39** (1989), 25-30.

[100] M. A. Khamsi, Etude de la propriete du point fixe dans les espaces metriques et les espaces de Banach, *These presentee a Paris 6*, Paris, 1987.

[101] M. A. Khamsi, On metric spaces with uniform normal structure, *Proc. Amer. Math. Soc.* **106** (1989), 723-726.

[102] M. A. Khamsi, On the weak*-fixed point property, *Contemp. Math. AMS* **85** (1989), 325-334.

[103] M. A. Khamsi, La propriete du point fixe dans les espaces de Banach avec base inconditionelle, *Math. Annalen* **277** (1987), 727-734.

[104] M. A. Khamsi, Normal structure for Banach spaces with Schauder decomposition, to apper in *Can. Math. Bull.*

[105] M. A. Khamsi, Points fixes de contractions dans les espaces de Banach, *Seminaire d'Initiation a l'Analyse* **86**, Univ. Paris VI, (1986-1987), 3-9.

[106] M. A. Khamsi, On normal structure, fixed point property and contraction of type (γ), *Proc. Amer. Math. Soc.* **106** (1989), 995-1001.

[107] M. A. Khamsi, W. M. Kozlowski, S. Reich, Fixed point theory in modular function spaces, to appear in *Nonlinear Analysis.*

[108] M. A. Khamsi, M. Pouzet, On 1-local retract of abstract sets and fixed point theory, to appear in *Proc. of Colloqui International Theorie Du. Fixe Marseille* (1989).

[109] M. A. Khamsi, S. Reich, Nonexpansive mappings and semigroups in hyperconvex spaces, to appear in *Math. Japonica.*

[110] M. A. Khamsi, Ph. Turpin, Fixed points of nonexpansive mappings in Banach lattices, *Proc. Amer. Math. Soc.* **105** (1989), 102-110.

[111] Y. Kijima, W. Takahashi, A fixed point theorem for nonexpansive mappings in metric spaces, *Kôdai Math. Sem. Rep.* **21** (1969), 326-330.

[112] W. A. Kirk, Fixed point theory for nonexpansive mappings I, *Lecture Notes in Math.* **886**, Springer-Verlag, Berlin (1981), 485-505.

[113] W. A. Kirk, Fixed point theory for nonexpansive mappings II, *Contemp. Math., Amer. Math. Soc.* **18** (1983), 121-140.

[114] W. A. Kirk, A fixed point theorem for mappings which do not increase distances, *Amer. Math. Monthly* **72** (1965), 1004-1006.

[115] W. A. Kirk, Fixed point theory in product spaces, Preprint.

[116] W. A. Kirk, Nonexpansive mappings in metric and Banach spaces, *Estratto Dai (Rendiconti del Seminaria Matematica e Fisico di Milano Vol. I* (1981), 133-144.

[117] W. A. Kirk, Nonexpansive mappings in product spaces, set valued mappings and k-uniform rotundity, *Nonlinear Func. Analysis, Proc. Symp. Pure Math., A.M.S.* **45** (1986), 52-64, Part 2.

[118] W. A. Kirk, The modulus of k-rotundity, Preprint.

[119] W. A. Kirk, Nonexpansive mappings and normal structure in Banach spaces, in *Proceedings of the Research Workshop on Banach Space Theory*, B. L. Lin, ed., University of Iowa, 1981.

[120] W. A. Kirk, Fixed point theorems in product spaces, operator equations and fixed points, *Math. Sci. Inst. Korea* **1** (1986), 27-35.

[121] W. A. Kirk, S. Massa, Remarks on asymptotic centers of sequences and nets, Preprint.

[122] W. A. Kirk, Y. Sternfield, The fixed point theory in certain product spaces, *Houston J. Math.* **10** (1984), 207-214.

[123] W. A. Kirk, C. M. Yanez, Nonexpansive and locally nonexpansive mappings in product spaces, Preprint.

[124] J. L. Krivine, B. Maurey, Espace de Banach stables, *Israel J. Math.* **39-40** (1981), 273-281.

[125] D. N. Kutzarova, S. L. Troyanski, Reflexive Banach spaces without equivalent norms which are uniformly convex or uniformly differentiable in every direction, *Stud. Math.* **72** (1982), 91-95.

[126] H. E. Lacey, *The isometric theory of classical Banach spaces*, Springer-Verlag **208**, 1974.

[127] E. Lami-Dozo, Centres asymptotiques dans certain F-espaces, *Bull. U.M.I. (5)* **17B** (1980), 740-747.

[128] E. Lami-Dozo, Operateurs nonexpansifs, P-compact et properietes geometriques de la norme, Ph.D. Thesis, Univ. De. Bruxelles, 1970.

[129] T. Landes, Permanence properties normal structure, *Pac. J. Math.* **100** (1984), 125-143.

[130] T. Landes, Normal structures and the sum-property, *Pac. J. Math. (1)* **123** (1986), 127-147.

[131] T. C. Lim, On the normal structure coefficients and the bounded sequence coefficients, *Proc. Amer. Math. Soc.* **88** (1983), 262-264.

[132] T. C. Lim, Asymptotic centers and nonexpansive mappings in conjugate Banach spaces, *Pac. J. Math.* **90** (1980), 135-143.

[133] T. C. Lim, Characterizations of normal structure, *Proc. of Am. Math. Soc.* **43** (1974), 313-319.

[134] M. Lin, R. Sine, Semigroups and retractions in hyperconvex spaces, Preprint.

[135] M. Lin, R. Sine, Contractive projections on the fixed point set of L_∞ contractions, Preprint.

[136] M. Lin, R. Sine, On the fixed point set of nonexpansive order preserving maps, to appear.

[137] P. K. Lin, Remarks on the fixed point problem for nonexpansive maps II, in *Texas Functional Analysis Seminar*, The Univ. of Texas, Austin, 1982-1983, 198-205.

[138] P. K. Lin, The Browder-Gohde property in product spaces, *Houston J. Math.* **13-2** (1987), 235-239.

[139] P. K. Lin, Unconditional bases and fixed points of nonexpansive mappings, *Pac. J. Math.* **116** (1985), 69-76.

[140] B. L. Lin, R. H. Lohman, On generalized James quasi reflexive Banach space, *Bull. Inst. Math. Acad. Sinica* **8** (1980), 389-399.

[141] J. Lindenstrauss, A. Pelcsynski, Absolutely summing operators in L_p-spaces and their applications, *Studia Math.* **29** (1968), 275-326.

[142] J. Lindenstrauss, H. P. Rosenthal, The L_p-spaces, *Israel J. Math.* **7** (1969), 325-347.

[143] J. Lindenstrauss, L. Tzafriri, Classical Banach spaces I, II, Springer-Verlag **92, 97**.

[144] R. Lipschitz, *Lehrbuch der Analyse*, Bonn, 1877.

[145] W. A. J. Luxemburg, A general theory of monads, in applications of model theory to algebra, analysis and probability, Holt, Rinehart and Winston, New York (1967), 18-85.

[146] H. V. Machado, Fixed point theorems for nonexpansive mappings in metric spaces with normal structure, Ph.D. Dissertation, University of Chicago, 1971.

[147] E. Maluta, Uniformly normal structure and related coefficients, *Pac. J. Math.* **111** (1984), 357-369.

[148] S. A. Mariadoss, P. M. Soardi, A remark on asymptotic normal

structure in Banach spaces, *Rend. Sem. Mat. Univ. Politech. Torino. (3)* **44** (1986), 393-395.

[149] B. Maurey, Points fixes des contractions sur un convex ferme de L^1, in *Seminaire d'Anlysis Fonctionnelle* **80-81**, Ecole Polytechnique, 1980.

[150] B. Maurey, G. Pisier, Series des variables aleatoires vectorielles independantes et proprietes geometriques des espaces de Banach, *Studia Math.* **58** (1976), 263-289.

[151] V. D. Milman, The geometric theory of Banach spaces, Part I, *Usp. Math. Nauk.* **25** (1970), 113-173 (Russian); English translation in *Russian Math. Surveys* **26** (1971), 79-163.

[152] J. L. Nelson, K. L. Singh, J. H. M. Whitfield, Normal structures and nonexpansive mappings in Banach spaces, in *Nonlinear Analysis*, World Sci. Publ., Singapore, 1987, 433-492.

[153] E. Odell, Y. Sternfeld, A fixed point theorem in c_0, *Pac. J. Math.* **95** (1981), 161-177.

[154] Z. Opial, Weak convergence of the sequence of succession approximations for nonexpansive mappings, *Bull. Amer. Math. Soc.* **73** (1967), 591-597.

[155] A. Pelczynski, On the impossibility of embedding the space L^1 in certain Banach spaces, *Coll. Math.* **8** (1961), 199-203.

[156] J. P. Penot, Fixed point theorem without convexity, *Analyse nonconvex* (1977, Pau) *Bull. Soc. Math. France*, Memoire 60 (1979), 129-152.

[157] E. Picard, Memoire sur la theorie des equations aux derives partielles et la methode des approximation successives, *J. de Math.* **6** (1890), 145-210.

[158] G. Pisier, Martingales with values in uniformly convex spaces, *Israel J. Math.* **20-3-4** (1975), 326-350.

[159] H. Poincare, Sur les courbes definies par les equations differentielles, *Jour. de Math.* **2**, 1886.

[160] A. Quilliot, Homomorphismes, point fixes, retraction et jeux de poursuite dans les graphes, les ensembles ordonnes et les espaces metriques, These de Doctorat D'état, Univ. Paris VI, 1983.

[161] W. A. Ray, R. C. Sine, Nonexpansive mappings with precompact orbits, in *Proceedings of the Workshop on Fixed Point Theory*, Univ. de Sherbrooke (1980), Lecture Notes in Math **886**, Springer-Verlag, New York (1981), 409-416.

[162] S. Reich, Integral equations, hyperconvex spaces, and the Hilbert ball, in *Nonlinear Analysis and Applications*, Marcel Dekker, New York (1987), 517-525.

[163] S. Reich, The fixed point property for nonexpansive mappings, *Amer. Math. Monthly* **83** (1976), 292-294.

[164] S. Reich, The fixed point property for nonexpansive mappings II, *Amer. Math. Monthly* **87** (1980), 292-294.

[165] H. P. Rosenthal, Weakly independent sequences and the Banach-Saks property, *Proc. of the Durham Symposium*, July 1975.

[166] H. P. Rosenthal, On subspaces of L_p, *Ann. of Math.* **97** (1973), 344-373.

[167] W. Rudin, *Functional Analysis*, McGraw-Hill, 1973.

[168] H. Scarf, The Approximation of fixed points of continuous mappings, *SIAM J. of Appl. Math.* **15** (1967), 1328-1343.

[169] J. Schauder, Zur theorie stetiger abbildungen in funktionalraumen, *Math. Z.* **26** (1927), 417-431.

[170] J. Schauder, Zur theorie stetiger abbildungen in funktionalraumen, *Math. Z.* **26** (1927), 47-65.

[171] J. Schauder, Der fixpunktsatz in funktionalraumen, *Stud. Math.* **2** (1930), 171-180.

[172] G. Schechtman, On commuting families of nonexpansive operators, *Proc. Amer. Math. Soc.* **84** (1982), 373-376.

[173] T. Sekowski, On normal structure, stability of fixed point property, and the modulus of noncompact convexity, *Inst. Matematyki U.M.C.S.*, 147-153.

[174] B. Sims, Ultra-techniques in Banach space theory, *Queen's Papers in Pure and Applied Math.* **60**, Ontario, 1982.

[175] R. C. Sine, On nonlinear contractions in sup norm spaces, *Nonlinear Analysis, TMA* **3** (1979), 885-890.

[176] R. C. Sine, Hyperconvexity and approximate fixed points, Preprint.

[177] R. C. Sine, Hyperconvexity and nonexpansive multifunctions, Preprint.

[178] R. C. Sine, Remarks on the example of Alspach, *Nonlinear Analysis and Applications*, Marcel Dekker (1982), 237-241.

[179] I. Singer, Bases in Banach spaces, **1, 2**, Springer, Berlin-Heidelberg-New York, 1970 and 1981.

[180] D. R. Smart, Fixed point theorems, *Cambridge Univ. Press,* 1974.

[181] M. A. Smith, Some examples concerning rotundity in Banach spaces, *Math. Ann.* **233** (1978), 151-161.

[182] M. A. Smith, B. Turett, Some examples concerning normal and uniform normal structure in Banach spaces, Preprint.

[183] M. A. Smith, B. Turett, Normal structure in Bocher L^p-spaces, to appear in *Pac. J. Math.*

[184] P. M. Soardi, Schauder bases and fixed point of non-expansive mappings, *Pac. J. Math.* **101** (1981), 193-198.

[185] P. M. Soardi, Existence of fixed point of non-expansive mappings in certain Banach lattices, *Proc. Amer. Math. Soc.* **73** (1979), 25-29.

[186] J. Stern, Propriétés locales et ultrapuissances d'espaces de Banach, *Seminaire Maurey-Schwartz,* 1974-1975, Ecole Polytechnique, Palaiseau, France.

[187] F. Sullivan, A generalization of uniformly rotund Banach spaces, *Can. J. Math. (3)* **31** (1979), 628-636.

[188] S. Swaminathan, Normal structures in Banach spaces and its generalizations, fixed points and nonexpansive mappings, *Contemp. Math., Amer. Math. Soc.* **18** (1983), 201-215.

[189] Y. Xin Tai, A geometrically aberrant Banach space with uniformly normal structure, Preprint.

[190] Y. Xin Tai, On k-UR spaces, *Contemp. Math, AMS Proc. of the Research Workshop on Banach Space Theory,* **85** (1987).

[191] Y. Xin Tai, D. Xingde, A fixed point theorem of asymptotically nonexpansive mappings, *J. of Math. (PRC) (3)* **6** (1986), 255-262.

[192] W. Takahashi, A convexity in metric space and nonexpansive mappings I, *Kôdai Math. Sem. Rep.* **22** (1970), 142-149.

[193] K. K. Tan, A note on asymptotic normal structure and close to normal structure, *Can. Math. Bull. (3)* **25** (1982), 339-343.

[194] B. Turret, A dual view of a theorem of Baillon, *Nonlinear Analysis and Applications,* **80** (1982), 279-286.

[195] B. S. Tzirelson, Not every Banach space contains ℓ_p or c_0, *Functional Anal. Appl.* **25** (1974), 138-141.

[196] C. M. Yanez, Geometric fixed point theory in Banach spaces, Ph.D. Thesis, Univ. of Iowa, August 1986.

[197] V. Zizler, On some rotundity and smoothness properties of Banach spaces, *Dissertationes Math.* **87** (1971), 5-33.

Index

A

approximate fixed point sequence (a.f.p.s.), 55
asymptotic normal structure, 54, 69

B

Banach contraction mapping principle, 49
Banach-Mazur distance, 31
Banach-Saks property (BSP), 38
basic sequence, 1
basis, 1
biorthogonal functional, 4
block basis, 2
boundedly complete basis, 4
bounded sequence coefficient, 68

C

characteristic of noncompact convexity, 66
characteristic of uniform convexity, 62
Chebyshev center, 53, 68
Chebyshev radius, 53
complete normal structure, 68
convexity structure, 69
convex type mapping, 86

E

equi-integrable function, 30

F

filter, 10
finite dimensional decomposition (F.D.D.), 8
finite representibility, 31
fixed point property, 50
Frechet filter, 10

G

G.L.D. property, 66

H

Hardy space, 97
hyperconvex, 70

J

James quasi-reflexive space, 78
James space, 46

K

Kuratowskii's measure of non-compactness, 66

L

Lipschitzian mapping, 40

M

modulus of convexity, 36

U

W